Analysis of The Global Water Crisis

Illustrating Regions with Extreme Water Scarcity to Those with Relative Abundance

RIVER PUBLISHERS SERIES IN SOCIAL, URBAN, ECONOMIC AND ENVIRONMENTAL SUSTAINABILITY

The River Series on Social, Urban, Economic and Environmental Sustainability is a series of comprehensive academic and professional books which focus on the societal side of sustainability. The series focuses on topics ranging from theory to policy and real-life case studies and applications.

Books published in the series include research monographs, edited volumes, handbooks and textbooks. The books provide professionals, researchers, educators, and advanced students in the field with an invaluable insight into the latest research and developments.

Topics included in this series are as follows:-

- Climate Change Mitigation
- Renewable Energy Policy
- Urban sustainability
- Strategic environmental planning
- Environmental Systems Monitoring and Analysis
- Greening the World Economy
- Sustainable Development

For a list of other books in this series, visit www.riverpublishers.com

Analysis of The Global Water Crisis

Illustrating Regions with Extreme Water Scarcity to Those with Relative Abundance

Medani P. Bhandari
Akamai University, USA,
Sumy State University, Ukraine,
and, Gandaki University, Nepal

River Publishers

Routledge
Taylor & Francis Group
NEW YORK AND LONDON

Published 2025 by River Publishers
River Publishers
Broagervej 10, 9260 Gistrup, Denmark
www.riverpublishers.com

Distributed exclusively by Routledge
605 Third Avenue, New York, NY 10158, USA
4 Park Square, Milton Park, Abingdon, Oxon OX14 4RN

Analysis of The Global Water Crisis: Illustrating Regions with Extreme Water Scarcity to Those with Relative Abundance / Medani P. Bhandari.

Routledge is an imprint of the Taylor & Francis Group, an informa business

ISBN 978-87-4380-105-4 (hardback)
ISBN 978-87-4380-189-4 (paperback)
ISBN 978-87-4380-107-8 (online)
ISBN 978-8-743-80106-1 (ebook master)

DOI: 10.1201/9788743801061

To my friend, wife, coauthor Mrs. Prajita Bhandari, without your help and support this book was impossible.

Together We Have Miles To Go

You have assumed a pivotal role akin to a true teacher, actively fostering an environmentally conscious atmosphere within your home, workplace, and community. Your impactful actions and exemplary behavior serve as an inspiration, motivating others to embrace sustainable practices and embrace a greener way of life. Your endeavors encompass a broad spectrum, from waste reduction to energy conservation and the advocacy for eco-friendly initiatives. Through these contributions, you actively create a better world for all. Your unwavering dedication to the stewardship of the environment shines as a guiding beacon, urging others to join the collective effort toward a healthier and more sustainable future. Amidst life's challenges, you provide me with a reassuring sense of comfort and a valuable lesson in resilience. For this, I extend my heartfelt gratitude to you.

Contents

Preface

Water is the essence of life, a fundamental resource that sustains ecosystems, communities, and economies. Yet, despite its critical importance, our planet faces a paradoxical crisis—regions abundant in freshwater resources coexist with areas suffering from extreme water scarcity. The global water crisis is a matter of availability but also of access, management, and equitable distribution. It is a complex, multifaceted issue that impacts nearly every aspect of human life, from health and food security to economic development and environmental sustainability. This book, *Analysis of The Global Water Crisis: Illustrating Regions with Extreme Water Scarcity to Those with Relative Abundance*, is an attempt to comprehensively explore this pressing challenge by examining various regions of the world, each with its unique water scenarios.

This work is the culmination of over four decades of my research, activism, and firsthand experiences with water scarcity. From a young age, I witnessed the hardships caused by inadequate access to clean drinking water in my community and various regions worldwide. These early experiences sparked a lifelong commitment to understanding and addressing the challenges related to water resources. Unfortunately, many issues I encountered as a child remain prevalent today, with billions of people still lacking access to safe and reliable water sources. The persistent nature of this crisis underscores the need for a deeper understanding and innovative solutions to ensure water security for all.

This book aims to offer a detailed overview of global freshwater distribution, highlighting the disparities between regions with extreme water scarcity and those with abundant water resources. The book is structured to provide a nuanced understanding of the factors influencing water availability and access, including climate variability, geographic conditions, population pressures, industrial demands, and infrastructure limitations. By categorizing the world into regions of extreme water scarcity, high to moderate water stress, moderate water security, and water abundance, this book seeks to illustrate the diversity of water challenges and opportunities across the globe.

In the chapters dedicated to Extreme Water Scarcity Regions, such as the Middle East and North Africa (MENA), South Asia, and Sub-Saharan Africa, I delve into the severe constraints these areas face due to limited freshwater resources, high population densities, and growing industrial and agricultural demands. These regions are often on the frontlines of the water crisis, experiencing acute shortages that threaten livelihoods, exacerbate poverty, and fuel conflicts over dwindling resources.

For High- to Moderate-Water-Stress Regions, like Central Asia, Southern Europe, the Mediterranean Basin, and parts of the Western United States, the focus shifts to the growing pressure on water resources due to over-extraction, pollution, and climate change. While not as severely constrained as the MENA region, these areas still face significant challenges in maintaining sustainable water supplies, necessitating innovative management strategies to prevent future crises.

In contrast, Regions with Moderate Water Security, including Eastern and Central Europe and Southeast Asia, demonstrate how effective management and infrastructural investments can mitigate water-related challenges. However, even these regions are not immune to the impacts of changing climate patterns and must remain vigilant in their efforts to secure water for future generations.

Finally, the exploration of Water-Abundant Regions such as South America, Russia, Canada, and Northern Europe underscores the unique challenges faced by areas with abundant water. While relatively secure in terms of water availability, these regions must address issues of environmental conservation, pollution control, and equitable distribution. The abundance of water does not necessarily translate to sustainable management, and there is a pressing need to protect these resources from the threats of deforestation, mining, urbanization, and industrial activities.

This book aims to bridge the knowledge gap by providing readers with a holistic understanding of the global water crisis, offering insights into both the challenges and solutions that can pave the way for a water-secure future. I hope this work will inform and inspire action among policymakers, researchers, practitioners, and concerned citizens committed to addressing the global freshwater crisis. The lessons drawn from diverse regional experiences can guide the development of tailored, region-specific strategies that ensure equitable access to water, promote sustainable use, and enhance resilience to future challenges.

By presenting a comprehensive analysis of global water distribution and the factors contributing to scarcity and abundance, I hope to open up new avenues for research, policy-making, and practical interventions that can mitigate the water crisis. I believe that with informed strategies, innovative technologies, and strong international cooperation, we can work towards a future where water scarcity is a problem of the past and water equity becomes a reality for all.

I dedicate this book to all those who have suffered and continue to suffer from the impacts of water scarcity and the countless individuals and organizations tirelessly working to solve one of the most critical challenges of our time. Let this book serve as a call to action to build a world where access to safe and sufficient water is a universal right, not a privilege.

Thank you!

Prof. Medani P. Bhandari, PhD.

Acknowledgements

The journey of writing "Analysis of The Global Water Crisis: Illustrating Regions with Extreme Water Scarcity to Those with Relative Abundance" has been one of the most profoundly fulfilling and deeply personal experiences of my life. This book is more than an academic endeavor; it embodies my lifelong commitment to exploring the intricate and fragile connections between humanity and the natural world. As I complete this project, I am overwhelmed with gratitude for the many individuals, experiences, and lessons that have profoundly influenced the content of this work and my perspective on the global water crisis.

This book is a culmination of decades of work as an environmental activist, humanitarian, and researcher. My career has been driven by an urgent call for sustainability, equity, and justice—issues that transcend disciplinary boundaries, national borders, and cultural divides. My ongoing mission has been to uncover ways we, as a global community, can coexist harmoniously with the natural world, ensuring its preservation for future generations. The path has been challenging but deeply rewarding, marked by significant reflections and a commitment to addressing the interconnected crises that threaten human and environmental well-being.

Throughout this journey, I have been fortunate to collaborate with a diverse range of people—from dedicated activists to scholars, policymakers, and community leaders. These encounters have underscored the interconnectedness of all life, offering invaluable lessons and inspiring me to persevere in advocating for sustainable change, even in the face of daunting global challenges. These collaborations have reinforced my belief in the power of collective action and shared knowledge.

Nature has also been a constant teacher, demonstrating resilience and balance within its ecosystems. Observing these natural phenomena has deepened my understanding of the vital relationship between human survival and the health of our environment. The insights gained from both human and ecological interactions are woven throughout this book, reflecting my belief that our future depends on finding holistic, inclusive solutions that prioritize sustainability and mutual flourishing for all life.

This book stands as a testament to the collective effort required to address the global freshwater crisis. It is my hope that it will inspire new perspectives and drive action towards a future where both people and nature can thrive in harmony. I extend my deepest gratitude to my family, especially my wife, Prajita, and daughter, Manaslu, whose unwavering support and encouragement have been the foundation of my journey. Their love and understanding have sustained me throughout this endeavor, making it a deeply personal and enriching experience.

Furthermore, I wish to acknowledge the countless individuals and communities I have encountered as an environmental conservationist and humanitarian. Your stories of resilience, compassion, and innovation have enriched my understanding of the core principles discussed in this book. I hope to honor your experiences by weaving them into the fabric of this narrative.

I sincerely thank my mentors and colleagues who have guided and challenged me throughout my career. Your insights, feedback, and collaborative spirit have greatly enriched my understanding and have been instrumental in developing this book. Special thanks to Prof. Steven R. Brechin, whose wisdom and encouragement have been invaluable, and to the academic institutions and communities, including Syracuse University, NY, Brandeis University, MA, USA, ITC, the Netherlands, and Tribhuvan University, Nepal, for the stimulating discussions and thoughtful critiques that have sharpened my perspective.

I am also grateful to the many scholars and researchers whose work has influenced and informed my own. The breadth and depth of knowledge in this field are vast, and the contributions of others have provided a rich tapestry of ideas upon which this book is built. To those whose writings and theories have inspired me, thank you for your intellectual generosity.

I extend my heartfelt gratitude to my dear friends, Mr. Kshitiz Raj Prasai, Rajan Adhikari, Dr. Medini Adhikari, Govinda Luitel, Binod Luitel and Prof. Bishwa Kalyan Parajuli. Your steadfast encouragement and insistence on continuing my research and writing have been a resolute driving force, bolstering my conviction in this meaningful endeavor.

I sincerely thank Prof. Mary Jo Bulbrook, Ph.D.; Prof. Ganesh Man Gurung, Ph.D.; Prof. Douglas Capogrossi, Ph.D.; Prof. Jacek Piotr Binda, Ph.D.; Dr. Ambika Adhikari, Ph.D.; Prof. Keshav Bhattarai, Ph.D.; Dr. Shvindina Hanna, Ph.D.; Dr. Aleksander Łukasz Sapi ski; Prof. Durga

Paudel, Ph.D.; Mr. Jay Bhandari, and others for their encouragement and support in my academic pursuits.

I extend my heartfelt gratitude to my mother, Heema Devi, whose uniqueness is a testament to the essence of all mothers, present and future, and Mother Earth herself. The depth of a mother's devotion is immeasurable, transcending enumeration. A mother's love remains matchless—until we fathom the depth of a mother's boundless, selfless affection, comprehending the very nature of maternal care, it is impossible to fully grasp the profound love and nurturing that Mother Earth extends to us and all living beings. Just as a mother's love is irreplaceable, so too is the immeasurable contribution of mother nature—a contribution that sustains and supports life in all its forms.

I would also like to extend my sincere gratitude to the invaluable contributions of my beloved family members, who have been unwavering pillars of support throughout my conservation journey. My wife, Prajita, along with my son, Prameya, daughter, Manaslu, and daughter-in-law, Kelsey Bhandari, son-in-law Abhimanyu, and his parents, Uma and Mahesh Iyer, thank you for your insightful comments and unwavering support during challenging times. To our grandchildren, Neena and Neelok, who brought joy to my family, I am grateful for your presence. I would also like to remember my two other grandchildren, Prasmin and Aahana Bhandari, and I hope that one day you will be reading this and other books of mine with pride.

My brothers Krishna and Shova, along with my nephews Jagannatha, Asmita, and Muna, brothers Hari and Durga, and daughter Pratigya, have been steadfast companions on this journey. My sisters Kali, Bhakti, Radha, Bindu, and Sita and their families have also provided unwavering support. Additionally, my uncle Bhim Prasad Bhandari and the late Goma Devi Bhandari, my siblings Govinda, Bishnu, Rudra, Bishma, and Sabita, together with their spouses and children, have been a constant source of strength and encouragement.

Prajita's parents, Dwarika Natha and Durga Devi, and her siblings Basudev, Ramesh, and Rishi, as well as Dhurba Acharya, Binod Ghimire, Ranjan Neupane, and Annapurna, Bijaya, Jaya, Ajaya, and their entire families have played an integral role in our endeavors. Without their unwavering support, my conservation journey would have faced interruptions or not have been possible. This sentiment applies universally—human beings thrive as social entities through mutual support.

A heartfelt and special acknowledgment is dedicated to my wife, Mrs. Prajita Bhandari, whose presence has created a tranquil space that enabled the completion of this book. Your perceptive insights and meticulous language editing have played an indispensable role in molding the outcome.

You have woven new threads into our journey—our endeavors in conservation and humanitarianism—and your support has been my strength throughout the trials of my life. Words are inadequate to convey my gratitude, and it is worth noting that this book is not just my narrative but also a testament to the journey we are embarking upon together.

I want to give a special thanks to River Publishers for their support and assistance in publishing this book. And to the future generations, whose well-being and future I hold close to my heart, thank you for motivating me to act and work towards a sustainable and thriving world for you.

Last but not least, I want to express my gratitude to all the readers. Your interest and engagement with my work fuels my passion for environmental conservation and social empowerment.

With a heart full of appreciation,

Thank you!

Prof. Medani P. Bhandari, PhD.

List of Tables

1

Analysis of The Global Water Crisis: Illustrating Regions with Extreme Water Scarcity to Abundance

Water is the foundation of life, essential for human survival, economic development, and ecosystem health. However, the world faces an alarming global water crisis characterized by severe scarcity in some regions and relative abundance in others. This crisis is not just a matter of natural availability but is deeply intertwined with issues of governance, infrastructure, socio-economic disparities, and climate change. While there are numerous scholarly books, research papers, and reports dedicated to analyzing water scarcity and management, there remains a significant knowledge gap in addressing the complexities of water governance, particularly the sustainable management of freshwater resources. This book, Analysis of the Global Water Crisis: Bridging Knowledge Gaps in Water Management seeks to fill these gaps by providing a comprehensive analysis of the challenges and solutions related to water management, combining theoretical frameworks with real-world case studies.

Water scarcity varies significantly across regions, with areas like the Middle East, North Africa, and parts of South Asia facing extreme shortages due to a combination of arid climates, high population densities, and over-extraction of groundwater. According to the United Nations, approximately 2 billion people currently live in countries experiencing high water stress, with this number projected to increase due to population growth, urbanization, and climate change (UN, 2021). Conversely, regions like Canada, Russia, and Brazil are rich in freshwater resources, yet these areas often face challenges related to water quality, pollution, and equitable distribution. This disparity in water availability raises critical questions about the global distribution of resources, social justice, and the role of effective water management policies. While several international frameworks, such as Integrated Water Resources Management (IWRM) and Sustainable Development Goal 6 (Clean Water and Sanitation), aim to address these issues, their implementation often

DOI: 10.1201/97818743801061-1

falls short due to a lack of localized solutions and stakeholder engagement (UN-Water, 2023).

Despite the extensive body of literature on water management, there are still significant knowledge gaps in understanding the socio-political dynamics that influence water scarcity, particularly in developing and conflict-prone regions. Much of the existing research tends to focus on technical solutions, such as the development of new technologies for water extraction and purification, without adequately addressing the socio-economic and political factors that drive mismanagement and inequitable access. For instance, studies have highlighted the role of corruption, inadequate infrastructure, and poor governance in exacerbating water crises in countries like Yemen, Syria, and India, yet there is limited research on how to effectively reform these systems (World Bank, 2018; Gleick, 2020). This book aims to bridge this knowledge gap by integrating insights from both technical and social sciences, offering a more holistic approach to water management that considers cultural, economic, and political dimensions.

Moreover, many existing studies are often region-specific or focused on a narrow set of issues, such as groundwater depletion or agricultural water use, without providing a comprehensive global perspective. This book seeks to provide a balanced analysis by exploring both the regions experiencing severe water scarcity and those with relative abundance, examining how policies, cultural practices, and socio-economic factors contribute to different outcomes. By highlighting case studies of both mismanaged and well-managed water systems—from the success of Singapore's water management strategy to the failures in the Aral Sea basin—this book aims to offer actionable insights for policymakers, practitioners, and researchers.

The motivation for writing this book stems from my personal and professional experiences with water scarcity. Growing up in a region where access to clean drinking water was often limited, I have been acutely aware of the critical importance of sustainable water management from an early age. Over the past four decades, I have dedicated my research and practical efforts to understanding the complexities of water governance, working across various regions, including Asia, Africa, and the Middle East, to address water-related challenges. This book is a culmination of these years of study and experience, aiming to provide new insights that can guide agents of change and stakeholders towards sustainable solutions.

One of the critical areas where this book seeks to contribute is in the development of a new framework for understanding and addressing the global water crisis. By combining theories from environmental economics, social justice, and resilience thinking with practical case studies, the book proposes a set of principles for sustainable water management that emphasizes the need for adaptive governance, community engagement, and equity. Unlike traditional approaches that often prioritize top-down solutions, this book advocates for a more decentralized, inclusive model that empowers local communities and integrates indigenous knowledge systems. This approach is particularly relevant in the context of climate change, where increasing variability in water availability necessitates more flexible and resilient management strategies (Rockström et al., 2009).

While the global water crisis is widely recognized, there remains a critical need for more integrated and holistic approaches to water management that go beyond technical fixes and address the underlying socio-political and economic drivers of scarcity Analysis of the Global Water Crisis: Bridging Knowledge Gaps in Water Management seeks to fill this gap by offering a comprehensive analysis of the global water crisis, drawing on four decades of research and practical experience. It is my hope that this book will serve as a valuable resource for researchers, policymakers, and practitioners, providing them with the tools and insights needed to foster sustainable water management practices that are both equitable and resilient.

1.1 Introduction

The global water crisis is a multifaceted issue that threatens the well-being of billions of people and poses significant risks to ecosystems worldwide. Despite water being a fundamental human right and a critical component for economic development, food security, and environmental sustainability, its scarcity and mismanagement have reached alarming levels. Today, over 2 billion people lack access to safe drinking water, while 4 billion experience severe water scarcity during at least one month of the year (UNICEF & WHO, 2021). This crisis is not just a result of physical water scarcity but also a product of poor governance, inadequate infrastructure, pollution, and socio-economic inequalities. Understanding the global water crisis requires a comprehensive analysis of regions facing extreme scarcity, areas of relative abundance, and the socio-political and environmental factors contributing to this imbalance.

1.2 The Scope of the Global Water Crisis

Water scarcity is not just an issue for arid and semi-arid regions but a global challenge that affects both developing and developed countries. According to the World Resources Institute (WRI), nearly one-quarter of the world's population lives in regions facing extremely high water stress, where demand exceeds supply (WRI, 2019). Countries like India, Pakistan, and Egypt are experiencing unprecedented levels of water stress due to a combination of rapid population growth, urbanization, and inefficient water use. For instance, the Ganges and Indus river basins, which support millions of livelihoods, are rapidly depleting due to over-extraction and pollution, threatening human communities and ecosystems (Jain et al., 2021).

In contrast, developed countries like the United States and Australia face challenges, particularly in regions like California and New South Wales, which are prone to severe droughts. Climate change increasingly impacts these areas, leading to prolonged periods of low rainfall, reduced snowpack, and shrinking reservoirs (Diffenbaugh et al., 2015). The crisis highlights that water scarcity is not merely a product of geography but also of governance and management practices, where even regions with abundant water resources can face shortages due to mismanagement.

1.3 Causes of the Global Water Crisis

A complex interplay of factors drives the global water crisis. Climate change disrupts rainfall patterns and intensifies droughts, while a growing population increases water demand. Pollution from industrial and agricultural activities contaminates water sources, rendering them unfit for consumption. Inefficient water use in agriculture and industry further exacerbates the issue. Poor water management practices, including inadequate infrastructure and lack of regulation, also contribute to water scarcity and pollution.

1.3.1 Climate change

As global temperatures rise, changes in precipitation patterns, melting glaciers, and increasing evaporation rates alter the distribution of freshwater resources. The Intergovernmental Panel on Climate Change (IPCC) predicts climate change will intensify the global hydrological cycle, leading to more frequent and severe droughts in some areas and increased flooding in others (IPCC, 2021).

Climate change is profoundly impacting the availability and distribution of freshwater resources globally. As global temperatures increase, the hydrological cycle—the continuous movement of water on, above, and below the surface of the Earth—is becoming more erratic. According to the Intergovernmental Panel on Climate Change (IPCC, 2021), rising temperatures accelerate the rate of evaporation, increase the atmosphere's capacity to hold moisture, and alter precipitation patterns. These shifts can result in severe droughts in some regions while causing intense flooding in others, thereby exacerbating water scarcity and affecting the livelihoods of millions of people around the world.

One of the most significant consequences of climate change on freshwater resources is the alteration of precipitation patterns. Many regions are experiencing changes in the frequency, intensity, and duration of rainfall, leading to either prolonged dry spells or extreme downpours. For example, the IPCC (2021) reports that areas such as the Mediterranean basin, southwestern United States, and parts of Africa are projected to experience reduced rainfall and increased drought conditions, while regions like Southeast Asia and parts of South America may face more intense rainfall events, resulting in frequent flooding (IPCC, 2021). Such shifts in precipitation not only affect the availability of surface water in rivers and lakes but also reduce groundwater recharge, further stressing water supplies in arid and semi-arid regions (Gleick, 2014).

Melting glaciers and snowpacks are another critical factor influencing freshwater availability. Glaciers in the Himalayas, Andes, and other mountainous regions serve as crucial sources of freshwater, especially during the dry seasons when meltwater supports rivers and agricultural systems downstream. However, studies indicate that many of these glaciers are shrinking rapidly due to rising global temperatures (Diffenbaugh, Swain, & Touma, 2015). The accelerated melting of glaciers can initially increase river flows, leading to seasonal flooding, but over the long term, it will significantly reduce the amount of freshwater available, especially for communities that rely heavily on glacial melt for their water needs (UNESCO, 2019). The depletion of these natural water reservoirs poses severe risks to food security, drinking water availability, and energy production in regions dependent on hydropower.

In addition, rising temperatures increase evaporation rates, particularly in already hot and dry areas, further diminishing water supplies in lakes, reservoirs, and soil moisture levels. For example, in regions like the American Southwest and parts of Australia, increased evaporation has contributed to significant reductions in surface water levels, leading to widespread water shortages and impacting agricultural productivity (Bakker & Cook, 2011). The IPCC (2021) highlights that climate-induced changes in evaporation and precipitation patterns will likely intensify competition for water resources among agricultural, industrial, and domestic users, particularly in water-stressed regions.

The impacts of climate change on freshwater resources also have socio-economic implications, including heightened risks of conflicts over water access and increased migration. As water becomes scarcer, competition among neighboring countries, regions, or communities over shared water bodies like rivers and aquifers could escalate, potentially leading to conflicts (Zawahri, 2017). Furthermore, populations in water-scarce areas may be forced to migrate to regions with more abundant water resources, thereby straining the existing infrastructure and resources of receiving areas (Sadoff & Grey, 2002). Addressing these challenges requires comprehensive water management strategies that incorporate climate resilience, sustainable water use practices, and international cooperation.

The effects of climate change on the global water cycle are multifaceted and far-reaching, with significant implications for water security, ecosystems, and human well-being. To mitigate these impacts, it is crucial to enhance adaptive measures such as improved water conservation techniques, sustainable agricultural practices, and investments in technologies like desalination and water recycling. The integration of these strategies into national and international water governance frameworks will be essential in ensuring a resilient and sustainable future for freshwater resources.

1.3.2 Population growth and urbanization: Increasing pressures on water resources

The world's population is expected to reach nearly 10 billion by 2050, significantly increasing demand for freshwater for drinking, agriculture, and industrial use (UN, 2019). Rapid urbanization, particularly in developing countries, places additional pressure on water infrastructure, often leading to over-extraction of groundwater and the contamination of surface water sources.

The rapid growth of the global population, projected to reach nearly 10 billion by 2050, is intensifying the demand for freshwater resources. As the United Nations (UN, 2019) highlights, this increase in population will significantly heighten the need for freshwater to support drinking water supplies, agriculture, and industrial processes. With more people comes the necessity to produce more food, which in turn requires substantial amounts of water for irrigation. The agricultural sector already consumes about 70% of global freshwater resources, and this figure is expected to rise as food production scales up to meet the needs of a growing population (FAO, 2019).

Rapid urbanization compounds these challenges, especially in developing countries where urban populations are expanding at unprecedented rates. The United Nations estimates that by 2050, nearly 70% of the world's population will reside in urban areas, up from 55% in 2018 (UN-Habitat,

2020). This urban expansion often outpaces the development of essential infrastructure, including water supply and sanitation systems, leading to severe water stress in many cities. Poorly planned urban growth can result in over-extraction of groundwater as cities strive to meet the water needs of their burgeoning populations. For instance, cities like Mexico City, Jakarta, and Dhaka face chronic water shortages due to over-reliance on rapidly depleting groundwater reserves (OECD, 2019).

Moreover, urbanization contributes to the contamination of surface water sources. As cities grow, so does the production of industrial waste, untreated sewage, and pollutants that often end up in rivers and lakes, making these water sources unsafe for human consumption. According to the World Health Organization (WHO, 2021), over 2 billion people worldwide lack access to safely managed drinking water services, and this issue is exacerbated in rapidly urbanizing regions where water treatment infrastructure is insufficient. The contamination of water bodies not only poses health risks but also reduces the availability of potable water, further straining limited freshwater resources (UNICEF & WHO, 2021).

The challenge of meeting the water demands of a growing and urbanizing population is particularly acute in regions already experiencing water scarcity. In parts of Africa, the Middle East, and South Asia, where water resources are already under significant stress due to arid climates and limited rainfall, population growth and urban sprawl are pushing these areas towards severe water crises (Zawahri, 2017). For example, the Middle East is one of the most water-scarce regions in the world, with some countries relying heavily on desalination to meet their water needs—a costly and energy-intensive solution (Elimelech & Phillip, 2011).

Addressing the dual challenges of population growth and urbanization requires a multifaceted approach, including investments in sustainable water management practices, the adoption of water-efficient technologies, and the strengthening of water governance frameworks. Efforts such as rainwater harvesting, wastewater recycling, and the implementation of more efficient irrigation techniques can help optimize water use in both urban and rural settings (Bakker, 2012). Furthermore, integrated urban water management strategies that combine infrastructure development with ecosystem conservation are essential for enhancing water security in rapidly growing cities (Gleick, 2014).

In conclusion, the pressures of a growing global population and accelerated urbanization are placing unprecedented demands on freshwater resources. Without strategic interventions, these trends threaten to exacerbate water scarcity, pollution, and socio-economic inequalities, particularly in regions already facing water stress. Sustainable water management, technological innovation, and improved governance are critical to ensuring that the world's expanding population can access the water it needs to thrive.

1.3.3 Agricultural demand: Unsustainable water use and its impacts

Agriculture is the largest consumer of freshwater, accounting for about 70% of global water withdrawals (FAO, 2020). The intensification of agriculture to meet the food demands of a growing population has led to unsustainable water use, particularly in water-scarce regions. Practices such as flood irrigation, inefficient water use, and the cultivation of water-intensive crops like rice and cotton exacerbate the crisis.

Agriculture remains the dominant consumer of global freshwater resources, accounting for approximately 70% of all water withdrawals (FAO, 2020). As the global population grows, there is increasing pressure to produce more food, which has led to the intensification of agricultural practices. However, this intensification often comes at a significant environmental cost, particularly in terms of water use. Many agricultural systems are characterized by inefficient practices, such as flood irrigation, which leads to significant water loss through evaporation and runoff (Gleick, 2014). Moreover, the cultivation of water-intensive crops, such as rice, sugarcane, and cotton, further exacerbates water scarcity, especially in arid and semi-arid regions where water resources are already stretched thin (FAO, 2019).

One of the critical challenges associated with agricultural water use is the over-extraction of groundwater. In regions like the Indo-Gangetic Plain, the Central Valley of California, and the North China Plain, the excessive pumping of groundwater for irrigation has led to significant declines in water tables, threatening long-term water availability (Wada et al., 2010). This unsustainable withdrawal of groundwater is particularly concerning in areas with limited natural recharge rates, where aquifers may take centuries to replenish. The Food and Agriculture Organization (FAO, 2020) warns that continued over-reliance on groundwater could lead to severe water shortages, impacting food security and livelihoods.

Additionally, the choice of crop types and irrigation methods plays a significant role in determining agricultural water use efficiency. For instance, growing water-intensive crops like rice and cotton in regions with limited water availability is often unsustainable. These crops require large amounts of water throughout their growth cycles, and in areas where water is already scarce, this can lead to further depletion of freshwater resources. For example, the cultivation of rice in water-stressed regions of India and Pakistan has significantly reduced groundwater levels, exacerbating water scarcity in these regions (Rodell et al., 2009). Similarly, the production of cotton in arid regions of Central Asia, such as Uzbekistan, has been linked to the shrinking of the Aral Sea due to extensive irrigation demands (Micklin, 2010).

Innovative approaches are needed to enhance water use efficiency in agriculture. Techniques such as drip irrigation, which delivers water directly to the root zone of plants, can reduce water consumption by 30-60% compared to traditional flood irrigation methods (Fereres & Soriano, 2007). Moreover, adopting drought-resistant crop varieties, implementing sustainable land management practices, and investing in technology for water recycling and reuse can mitigate the impact of agriculture on freshwater resources (Rockström et al., 2010). Such measures are essential to ensure that agricultural practices are sustainable and that they do not compromise the water needs of future generations.

In conclusion, while agriculture is vital for food security, its intensive water use presents significant challenges for sustainable water management. Addressing these challenges requires a combination of policy reforms, technological innovations, and changes in agricultural practices to optimize water use and protect freshwater resources.

1.3.4 Pollution: A significant threat to freshwater resources

Industrial discharges, agricultural runoff, and untreated sewage contribute to the pollution of freshwater sources, making water unsafe for human consumption and damaging aquatic ecosystems. For example, the overuse of fertilizers and pesticides has led to the contamination of rivers and lakes, resulting in algal blooms and the destruction of aquatic life (Gleick, 2014).

Pollution is one of the most pervasive threats to freshwater resources worldwide. Industrial discharges, agricultural runoff, and untreated sewage are major contributors to the contamination of rivers, lakes, and groundwater, rendering water unsafe for human consumption and harming aquatic ecosystems. As industries expand, the discharge of toxic chemicals, heavy metals, and other pollutants into water bodies has escalated, often without adequate treatment. For example, factories frequently release effluents containing hazardous substances like mercury, lead, and arsenic into nearby rivers, which can accumulate in the food chain, posing severe health risks to both humans and wildlife (Schwarzenbach et al., 2010).

Agricultural activities are another significant source of freshwater pollution. The extensive use of chemical fertilizers and pesticides in modern farming has resulted in nutrient pollution, especially nitrogen and phosphorus, which leach into rivers and lakes. This nutrient loading stimulates the growth of algal blooms, which deplete oxygen levels in the water, causing dead zones where aquatic life cannot survive (Carpenter et al., 1998). In severe cases, algal blooms produce toxins that can contaminate drinking water supplies, as seen in incidents affecting large freshwater lakes like Lake Erie and the Baltic Sea (Michalak et al., 2013).

Urbanization further exacerbates water pollution through untreated sewage and stormwater runoff. In many developing countries, inadequate infrastructure for wastewater treatment leads to the direct discharge of untreated sewage into water bodies, contributing to the spread of waterborne diseases like cholera, dysentery, and typhoid (UNICEF & WHO, 2021). This problem is particularly acute in densely populated urban areas, where rapid growth often outstrips the capacity of municipal water treatment facilities, leading to severe contamination of local water sources.

Moreover, plastic pollution has emerged as a growing threat to freshwater ecosystems. Microplastics, tiny plastic particles originating from the breakdown of larger plastic waste or microbeads in personal care products, have been found in rivers, lakes, and drinking water sources (Pivokonsky et al., 2018). These microplastics can absorb toxic chemicals and serve as pollutant vectors, potentially impacting aquatic organisms and human health.

Addressing the issue of water pollution requires a multi-faceted approach. Strategies include strengthening regulations on industrial discharges, promoting sustainable agricultural practices that reduce nutrient runoff, and investing in improved wastewater treatment infrastructure (OECD, 2019). Public awareness campaigns and community engagement are essential to promote better waste management practices and reduce pollution at the source.

1.3.5 Poor governance and inequality: Barriers to equitable water access

The global water crisis is not just about physical scarcity but also about access and equity. Many regions lack the infrastructure and governance frameworks to manage water resources sustainably. Water management often suffers from fragmentation, corruption, and a lack of coordination among stakeholders, leading to inefficient use and wastage (Bakker, 2012). Moreover, marginalized communities often face discrimination in access to clean water, exacerbating socio-economic disparities.

The global water crisis extends beyond mere physical scarcity; it is deeply intertwined with issues of governance, access, and social equity. While some regions face severe natural limitations in water availability, others suffer from mismanagement and inequitable distribution of existing resources. Poor governance is a significant factor exacerbating the water crisis, as it often results in inefficient water use, wastage, and a failure to sustainably manage water resources (Bakker, 2012; Sadoff & Grey, 2002).

Fragmentation in water management systems is a critical governance issue. Many countries lack cohesive frameworks that coordinate water use across sectors such as agriculture, industry, and domestic needs. Instead, water management is often divided among multiple agencies with overlapping responsibilities and conflicting interests, leading to inefficient and unsustainable outcomes (Bakker & Cook, 2011). This fragmentation is further complicated by corruption, weak regulatory mechanisms, and inadequate enforcement of water laws, which undermine efforts to protect and allocate water resources effectively (OECD, 2019). For instance, in many developing countries, the absence of transparent governance structures allows powerful stakeholders to dominate water access, leaving marginalized communities with insufficient supplies (Gleick, 2014).

Inequity in water access is another pressing challenge linked to governance failures. Marginalized communities, particularly in developing regions, often face systemic discrimination that limits their access to clean and safe water. This disparity is evident in rural areas, where inadequate infrastructure prevents effective water distribution, and in urban slums, where informal settlements are typically excluded from municipal water networks (UNICEF & WHO, 2021). Moreover, the privatization of water services in some regions has led to increased water costs, making it unaffordable for low-income households, further deepening socio-economic inequalities (Bakker, 2012).

The implications of poor water governance are also visible on a global scale. Transboundary water conflicts, where rivers and lakes span multiple countries, highlight the need for international cooperation. Without effective governance frameworks, these shared water resources become sources of tension, as seen in regions like the Middle East and North Africa, where water scarcity and political instability intersect (Zawahri, 2017).

Addressing the governance and inequality dimensions of the water crisis requires a shift towards more inclusive and integrated water management strategies. Solutions include enhancing regulatory frameworks, promoting stakeholder participation, and ensuring equitable access to water for all communities, regardless of their socio-economic status. By strengthening governance structures, countries can move towards sustainable water management that not only addresses scarcity but also upholds the principles of social justice and equity (Sadoff & Grey, 2002; Rockström et al., 2014).

1.4 Regional Case Studies: From Scarcity to Abundance

Water scarcity is a pressing global issue, particularly in arid and semi-arid regions. Countries like India and China face significant water stress with their rapidly growing populations and intensive agriculture. Limited water resources and increasing demand have led to conflicts and tensions in the Middle East and North Africa. Climate change further exacerbates the problem, altering rainfall patterns and intensifying droughts. These regions are grappling with water shortages for domestic use, agriculture, and industry, highlighting the urgent need for sustainable water management practices.

1.4.1 Extreme scarcity in the Middle East and North Africa (MENA)

The MENA region is the most water-scarce region in the world, with many countries relying heavily on non-renewable groundwater and desalination to meet their water needs. Saudi Arabia, for example, has turned to large-scale desalination projects to supply nearly 70% of its drinking water (World Bank, 2020). However, this solution is energy-intensive and not sustainable in the long term, particularly in the face of rising energy costs and environmental concerns. Conflicts over shared water resources, such as the Nile, Jordan, and Euphrates rivers, further complicate water security in this region, often leading to political tensions (Sadoff & Grey, 2002).

1.4.2 The paradox of water abundance in Sub-Saharan Africa

Sub-Saharan Africa faces a unique challenge where, despite having major river systems and substantial rainfall in some areas, millions of people lack access to clean water due to inadequate infrastructure and governance. Countries like Nigeria and the Democratic Republic of Congo are rich in water resources but struggle with poor water management, leading to severe water-related health issues like cholera outbreaks (UNICEF, 2021). Investments in water, sanitation, and hygiene (WASH) infrastructure are essential to bridge this gap and improve health outcomes in the region.

1.4.3 Water abundance in Canada and Scandinavia

Countries like Canada and Norway enjoy abundant freshwater resources due to their extensive river systems and glacial reserves. Canada, for instance, has around 20% of the world's freshwater, mostly located in its lakes and rivers (Environment Canada, 2018). However, even these water-rich countries face challenges related to water quality, indigenous water rights, and the impact of climate change on water resources (Bakker & Cook, 2011). Effectively managing water resources in these regions highlights the importance of strong governance and integrated water management approaches.

1.5 Potential Solutions and Pathways Forward

Addressing the global water crisis requires a combination of technological, policy, and community-based solutions. Key strategies include:

1.5.1 Sustainable water management

Implementing integrated water resource management (IWRM) approaches that balance water use across sectors can improve efficiency and sustainability. This involves the use of smart irrigation technologies, rainwater harvesting, and wastewater recycling (Elimelech & Phillip, 2011).

1.5.2 Climate adaptation strategies

Enhancing the resilience of water systems to climate change is crucial. This includes building infrastructure for water storage, flood control, drought mitigation, and restoring natural ecosystems like wetlands that provide water regulation services (Rockström et al., 2014).

1.5.3 Strengthening water governance

Reforming water policies to ensure equitable access, transparency, and community involvement in decision-making is vital. This can help prevent conflicts over water resources and promote sustainable use (OECD, 2019).

1.5.4 Investing in innovation

Desalination technologies, though costly, offer a lifeline for water-scarce regions. Advances in renewable energy can help reduce the environmental impact of desalination. Additionally, emerging technologies like atmospheric water generators and water-efficient crops could provide innovative solutions to water scarcity (Vörösmarty et al., 2010).

The global water crisis underscores the urgent need for coordinated action at the local, national, and international levels. By addressing both the physical and socio-economic dimensions of water scarcity, sustainable solutions can be created that ensure water security for future generations. A holistic approach that integrates technological innovation, policy reform, and community engagement is essential for overcoming the challenges of water scarcity and achieving the Sustainable Development Goal of clean water and sanitation for all.

This analysis demonstrates that while the global water crisis is complex, it is not insurmountable. By understanding the underlying causes and implementing strategic solutions, we can achieve a more sustainable and water-secure future.

1.6 Analysis of The Global Water Crisis: A Global Perspective from Scarcity to Abundance

The global water crisis represents one of the most pressing challenges of the 21st century, characterized by an alarming imbalance in water availability, access, and distribution. Despite water covering approximately 71% of the Earth's surface, only 2.5% of this is freshwater, with the majority locked in glaciers, ice caps, or deep underground. The small fraction of accessible freshwater is further strained by increasing demand, pollution, climate change, and socio-economic inequalities, creating a complex web of challenges that threaten human well-being and ecological systems (Gleick, 2014). This analysis explores the global water crisis by examining regions facing extreme water scarcity and contrasting them with areas of relative abundance, highlighting the factors contributing to these disparities.

1.6.1 Regions with extreme water scarcity: The case of the Middle East and North Africa (MENA)

The Middle East and North Africa (MENA) region is often cited as one of the most water-scarce areas in the world, characterized by arid and semi-arid climates with limited freshwater resources. Countries like Yemen, Saudi Arabia, and Jordan face acute water shortages, relying heavily on groundwater extraction and desalination to meet their needs. For instance, Yemen's per capita water availability is less than 100 cubic meters per year, far below the international water scarcity threshold of 1,000 cubic meters (World Bank, 2020). The region's dependence on non-renewable groundwater resources is unsustainable, with over-extraction leading to the depletion of aquifers, land subsidence, and increased salinity.

The socio-political landscape further exacerbates water scarcity in the MENA region. Conflicts over transboundary water resources, such as the

Nile, Tigris, and Euphrates rivers, have intensified tensions between riparian states (Zawahri, 2017). Additionally, the MENA region's rapid urbanization, population growth, and climate change are putting additional pressure on already scarce water supplies. The Intergovernmental Panel on Climate Change (IPCC) projects that the MENA region will experience a decrease in precipitation and an increase in temperature, leading to more frequent droughts and reduced water availability (IPCC, 2021).

1.6.2 Sub-Saharan Africa: A dual challenge of scarcity and poor infrastructure

Sub-Saharan Africa presents a paradox of water abundance in some areas and severe scarcity in others. While the region is endowed with major river systems like the Congo, Niger, and Zambezi, many countries struggle with water access due to inadequate infrastructure, poor governance, and socio-economic challenges. Countries such as Ethiopia, Somalia, and Sudan are prone to severe droughts, which are becoming more frequent due to climate change. The Horn of Africa, for example, has faced prolonged drought periods, resulting in food insecurity, displacement, and conflict over dwindling resources (FAO, 2019).

Even in areas where water is relatively abundant, the need for more infrastructure for water storage, treatment, and distribution limits access. Over 400 million people in Sub-Saharan Africa lack access to basic drinking water services, with rural communities being the most affected (UNICEF, 2020). The region's struggle with water-borne diseases, such as cholera and typhoid, further underscores the need for investments in water, sanitation, and hygiene (WASH) infrastructure to improve health outcomes.

1.6.3 Contrasting abundance: The case of Canada and the Nordic Countries

Countries like Canada, Norway, and Sweden have abundant freshwater resources, with vast lakes, rivers, and groundwater reserves compared to water-scarce regions. Canada alone holds about 20% of the world's freshwater, with its Great Lakes system representing Earth's largest group of freshwater lakes (Environment Canada, 2018). The Nordic countries, known for their efficient water management systems, have leveraged their abundant resources to achieve high living standards with universal access to clean drinking water and sanitation.

The abundance of water in these regions is not merely a result of natural endowments but also effective water governance. Canada, for example, has implemented integrated watershed management approaches, emphasizing the sustainable use of water resources, conservation, and protection of aquatic ecosystems (Bakker & Cook, 2011). Similarly, the Nordic water management model prioritizes environmental sustainability, with policies that protect water bodies from pollution, sustainable agriculture practices, and the promotion of green infrastructure (OECD, 2019).

1.6.4 Emerging solutions and policy implications

Addressing the global water crisis requires a multifaceted approach that combines technological innovations, policy reforms, and community engagement. Innovations such as desalination, water recycling, and rainwater harvesting offer potential solutions to augment water supplies, particularly in arid regions (Elimelech & Phillip, 2011). However, these technologies are often costly and require significant energy inputs, which may not be feasible for low-income countries.

Policy interventions must prioritize equitable access to water resources, emphasizing the human right to water as recognized by the United Nations General Assembly (UN, 2010). Strengthening transboundary water governance frameworks, particularly in conflict-prone areas, is crucial for fostering cooperation and preventing water-related conflicts (Sadoff & Grey, 2002). Additionally, investing in water conservation, storage, and distribution infrastructure, particularly in Sub-Saharan Africa, can help bridge the gap between water abundance and access.

The global water crisis manifests natural and human-induced factors, revealing stark disparities between regions facing extreme scarcity and those enjoying relative abundance. As climate change continues to alter precipitation patterns and exacerbate water stress, sustainable water management becomes even more critical. A holistic approach that combines technological, policy and community-based solutions is essential to address the water crisis and ensure that water security is achieved for all.

References

Bakker, K. (2012). Water security: Research challenges and opportunities. *Science, 337*(6097), 914–915. https://doi.org/10.1126/science.1226337

Bakker, K., & Cook, C. (2011). Water Governance in Canada: Innovation and Fragmentation. *Water Resources Development, 27*(2), 275–289. https://doi.org/10.1080/07900627.2011.564969

Carpenter, S. R., Caraco, N. F., Correll, D. L., Howarth, R. W., Sharpley, A. N., & Smith, V. H. (1998). Nonpoint pollution of surface waters with phosphorus and nitrogen. *Ecological Applications, 8*(3), 559–568. https://doi.org/10.1890/1051-0761(1998)008[0559:NPOSWW]2.0.CO;2

Diffenbaugh, N. S., Swain, D. L., & Touma, D. (2015). Anthropogenic warming has increased drought risk in California. *Proceedings of the National Academy of Sciences, 112*(13), 3931–3936. https://doi.org/10.1073/pnas.1422385112

Elimelech, M., & Phillip, W. A. (2011). The future of seawater desalination: Energy, technology, and the environment. *Science, 333*(6043), 712–717. https://doi.org/10.1126/science.1200488

FAO. (2019). Drought in the Horn of Africa: An urgent need for collective action. Food and Agriculture Organization of the United Nations. FAO Report

FAO. (2020). The state of food and agriculture: Overcoming water challenges in agriculture. Food and Agriculture Organization of the United Nations. FAO Report

Fereres, E., & Soriano, M. A. (2007). Deficit irrigation for reducing agricultural water use. *Journal of Experimental Botany, 58*(2), 147–159. https://doi.org/10.1093/jxb/erl165

Gleick, P. H. (2014). Water, drought, climate change, and conflict in Syria. *Weather, Climate, and Society, 6*(3), 331–340. https://doi.org/10.1175/WCAS-D-13-00059.1

Gleick, P. H. (2020). *The World's Water: The Biennial Report on Freshwater Resources*. Island Press.

IPCC. (2021). *Climate change 2021: The physical science basis*. Intergovernmental Panel on Climate Change. https://www.ipcc.ch/report/ar6/wg1/

Michalak, A. M., Anderson, E. J., Beletsky, D., Boland, S., Bosch, N. S., Bridgeman, T. B., ... & Zagorski, M. A. (2013). Record-setting algal bloom in Lake Erie caused by agricultural and meteorological trends consistent with expected future conditions. *Proceedings of the National Academy of Sciences, 110*(16), 6448–6452. https://doi.org/10.1073/pnas.1216006110

Micklin, P. (2010). The past, present, and future of the Aral Sea. *Lake and Reservoir Management, 25*(3), 1–15. https://doi.org/10.1080/07438140903105916

OECD. (2019). Water governance in the Nordic countries: Key insights and lessons learned. Organisation for Economic Co-operation and Development. OECD Report

OECD. (2019). *Water governance in the Nordic countries: Key insights and lessons learned*. Organisation for Economic Co-operation and Development. https://doi.org/10.1787/9789264315326-en

OECD. (2019). Water Governance in the OECD: Progress on Policy Reforms. *Organization for Economic Co-operation and Development*. https://doi.org/10.1787/8f10f4cd-en

Pivokonsky, M., Cermakova, L., Novotna, K., Peer, P., Cajthaml, T., & Janda, V. (2018). Occurrence of microplastics in raw and treated drinking water. *Science of The Total Environment, 643*, 1644–1651. https://doi.org/10.1016/j.scitotenv.2018.08.102

Rockström, J., et al. (2010). Managing water in rainfed agriculture—The need for a paradigm shift. *Agricultural Water Management, 97*(4), 543–550. https://doi.org/10.1016/j.agwat.2009.09.009

Rockström, J., et al. (2014). *Water resilience for human prosperity*. Cambridge University Press. https://doi.org/10.1017/CBO9781316091554

Rockström, J., Steffen, W., Noone, K., et al. (2009). "A Safe Operating Space for Humanity." *Nature*, 461(7263), 472–475.

Rodell, M., Velicogna, I., & Famiglietti, J. S. (2009). Satellite-based estimates of groundwater depletion in India. *Nature, 460*(7258), 999–1002. https://doi.org/10.1038/nature08238

Sadoff, C. W., & Grey, D. (2002). Beyond the River: The Benefits of Cooperation on International Rivers. *Water Policy, 4*(5), 389–403. https://doi.org/10.1016/S1366-7017(02)00035-1

Schwarzenbach, R. P., Escher, B. I., Fenner, K., Hofstetter, T. B., Johnson, C. A., Von Gunten, U., & Wehrli, B. (2010). The challenge of micropollutants in aquatic systems. *Science, 313*(5790), 1072–1077. https://doi.org/10.1126/science.1127291

UN. (2019). World population prospects 2019: Highlights. United Nations. UN Report

UN. (2021). *The State of Food Security and Nutrition in the World*. United Nations.

UNESCO. (2019). *The United Nations World Water Development Report 2019: Leaving No One Behind*. United Nations Educational, Scientific and Cultural Organization. https://en.unesco.org/themes/water-security/wwap/wwdr/2019

UN-Habitat. (2020). World cities report 2020: The value of sustainable urbanization. United Nations Human Settlements Programme. UN-Habitat Report

UNICEF & WHO. (2021). Progress on Household Drinking Water, Sanitation, and Hygiene 2000-2020: Five Years into the SDGs. United Nations Children's Fund and World Health Organization. UNICEF & WHO Report

UNICEF & WHO. (2021). Progress on household drinking water, sanitation, and hygiene 2000-2020: Five years into the SDGs. UNICEF Report

UNICEF. (2020). *Progress on drinking water, sanitation, and hygiene: 2019 update and SDG baselines*. United Nations Children's Fund. https://data.unicef.org/resources/progress-drinking-water-sanitation-hygiene-2019-update-sdg-baselines/

UN-Water. (2023). *Progress on Water-Related Sustainable Development Goals*. United Nations.

Wada, Y., Van Beek, L. P. H., & Bierkens, M. F. P. (2010) Modelling global water stress of the recent past: On the relative importance of trends in water demand and climate variability. *Hydrology and Earth System Sciences, 14*(10), 2201–2219. https://doi.org/10.5194/hess-14-2201-2010

WHO. (2021). Drinking-water fact sheet. World Health Organization. WHO Fact Sheet

World Bank. (2018). *Beyond Scarcity: Water Security in the Middle East and North Africa*. World Bank Publications.

World Bank. (2020). *Water in the Arab World: Management perspectives and innovations*. https://documents.worldbank.org/en/publication/documents-reports/documentdetail/595601468739449421/water-in-the-arab-world-management-perspectives-and-innovations

World Resources Institute (WRI). (2019). *Aqueduct water risk atlas*. https://www.wri.org/aqueduct

Zawahri, N. A. (2017). The politics of water scarcity in the Middle East. *International Studies Review, 19*(2), 258–280. https://doi.org/10.1093/isr/vix012

2

Analysis of The Global Water Crisis: Introduction to Global Water Distribution

2.1 Introduction to Global Water Distribution

Water is the most critical natural resource for sustaining life on Earth, yet its global distribution is highly uneven. Although approximately 71% of the Earth's surface is covered with water, only about 2.5% of this is freshwater, which is suitable for human consumption, agriculture, and industrial use (Shiklomanov, 1993). The vast majority of the planet's water, about 97.5%, is saline and contained in oceans, making it unsuitable for most human needs without costly desalination processes. Of the limited freshwater resources, nearly 70% is locked in glaciers and ice caps, primarily in Antarctica and Greenland, leaving less than 1% of the world's total water resources accessible in rivers, lakes, and underground aquifers (Gleick, 2019). This unequal distribution highlights the challenges faced by many regions in accessing sufficient freshwater to meet their needs, particularly in arid and semi-arid areas.

The geographic distribution of freshwater is not only uneven but also subject to significant temporal variability due to seasonal changes, climate patterns, and human activities. For example, tropical regions like the Amazon basin receive high levels of precipitation, resulting in abundant freshwater resources, whereas regions like the Middle East and North Africa experience chronic water scarcity due to their arid climates (Mekonnen & Hoekstra, 2016). The situation is further complicated by the fact that water availability is influenced by the interplay between natural factors such as rainfall and human-induced factors like population growth, urbanization, pollution, and inefficient water management practices (Postel, 2017). For instance, despite having significant water resources, countries like India and China face severe water stress due to the over-extraction of groundwater and the pollution of surface waters, driven by rapid industrialization and agricultural demands (WWF, 2020).

DOI: 10.1201/9781874380161-2

The global water distribution is also impacted by the hydrological cycle, which governs the movement of water between the atmosphere, land, and oceans. This cycle is critical for replenishing freshwater sources through processes like precipitation, infiltration, and runoff. However, human activities such as deforestation, land use change, and climate change have disrupted the natural hydrological balance, leading to altered precipitation patterns, reduced groundwater recharge, and increased incidences of droughts and floods (IPCC, 2021). For example, the over-extraction of groundwater for irrigation in regions like the High Plains Aquifer in the United States and the North China Plain has led to significant declines in water tables, threatening the long-term sustainability of these critical water resources (Scanlon et al., 2012). Such disruptions are not only reducing the availability of freshwater but also contributing to the degradation of ecosystems that depend on these water sources.

Socio-economic factors further exacerbate the inequitable distribution of water. High-income countries often have the resources to develop infrastructure for water storage, treatment, and distribution, whereas low-income countries may lack the capacity to invest in such systems, leading to disparities in water access (UNESCO, 2020). Additionally, transboundary water bodies like rivers and aquifers that cross national borders are sources of both cooperation and conflict. The Nile, Ganges, and Jordan rivers, for instance, are shared by multiple countries, leading to complex dynamics of water allocation and rights (Wolf, 2007). Effective management of these shared resources requires international cooperation, yet political tensions and competing national interests often hinder collaborative efforts.

Given the complexity of global water distribution, it is essential to adopt a comprehensive and integrated approach to water management that considers both natural and human-induced factors. This book aims to provide a deeper understanding of global water distribution patterns, emphasizing the need for sustainable and equitable water management practices. By examining case studies from regions with abundant water resources, such as Canada and Brazil, as well as those facing severe scarcity, like the Middle East and Sub-Saharan Africa, this book seeks to highlight best practices and lessons learned. These insights can inform strategies to enhance water security, improve governance, and promote the sustainable use of freshwater resources in the face of growing global challenges.

Despite extensive research on water resources, significant gaps remain in our understanding of how best to manage and distribute freshwater in a way that ensures sustainability and equity. Existing studies often focus on the technical aspects of water management, such as infrastructure development and technological innovations, without adequately addressing the socio-economic and political dimensions that are critical to sustainable water use (Rogers & Hall, 2003). This book aims to fill these gaps by providing a holistic analysis that integrates scientific, socio-economic, and policy perspectives. It draws on over four decades of research and practical experience to offer new frameworks and strategies for addressing the global water crisis, with a focus on inclusive, community-based approaches that prioritize social equity and environmental sustainability.

2.2 Overview of Global Freshwater Availability

Freshwater is a finite and precious resource that is crucial for human survival, agriculture, industry, and ecosystems. Despite the Earth's surface being covered by approximately 71% water, only about 2.5% of this is freshwater, and even less is accessible for use by humans (Shiklomanov, 1993). Of this 2.5%, nearly 70% is locked away in glaciers and ice caps, primarily in Antarctica and Greenland, leaving a mere 1% available in rivers, lakes, and groundwater for all human and ecological needs (Gleick, 2019). These accessible freshwater resources are unevenly distributed across the globe, resulting in significant regional disparities in water availability. While some regions are rich in freshwater resources, others face extreme scarcity, and these imbalances are exacerbated by population growth, urbanization, climate change, and unsustainable water management practices.

2.2.1 Distribution of freshwater resources

The distribution of freshwater across the world is heavily influenced by geographical, climatic, and hydrological factors. Some areas of the world, such as Canada, Brazil, and Russia, possess vast quantities of freshwater in the form of lakes, rivers, and groundwater, which are more than sufficient to meet their domestic and agricultural demands (Gleick, 1993). For instance, Canada holds approximately 20% of the world's renewable freshwater supply, largely due to the vast expanses of freshwater lakes and rivers within its territory (Postel, 2017). Similarly, the Amazon Basin in Brazil harbors a significant portion of the world's freshwater in the form of river systems, making the country a key player in global freshwater distribution. However, even within these water-rich regions, water quality and distribution across different regions can present challenges, especially as pollution and regional imbalances in water use arise.

On the other hand, many parts of the world face severe freshwater scarcity, where water availability per capita is significantly lower than the global average. The Middle East and North Africa (MENA) region is the most water-scarce region globally, with some countries like Saudi Arabia, Jordan, and Yemen having access to less than 100 cubic meters of freshwater per person annually, far below the threshold for water stress, which is generally considered to be 1,000 cubic meters per capita (Mekonnen & Hoekstra, 2016). The arid climate and over-reliance on non-renewable groundwater resources have exacerbated water scarcity in this region, leading to unsustainable water extraction practices and an over-dependence on desalinated water. Similarly, Sub-Saharan Africa faces challenges with water access, with many countries experiencing chronic water shortages due to high population growth rates and the increasing demand for agricultural water (United Nations, 2018).

2.2.2 Groundwater: A vital source of freshwater

Groundwater is a critical source of freshwater, especially in arid regions and for agricultural activities. It accounts for about 30.1% of the world's total freshwater supply, with the remainder being surface water stored in lakes, rivers, and reservoirs (Shiklomanov, 1993). Groundwater is often considered a more reliable source of freshwater since it is less vulnerable to seasonal fluctuations and is not directly impacted by surface water pollution. However, the over-extraction of groundwater has led to significant declines in water tables in many regions. For example, in South Asia, particularly in India and Pakistan, the excessive use of groundwater for irrigation has led to widespread depletion, threatening the long-term sustainability of water resources (Bierkens, 2015). This trend has raised concerns about the availability of groundwater for future generations, especially in regions where alternative water sources are scarce.

2.2.3 Climate change and freshwater availability

Climate change is another critical factor influencing global freshwater availability. The increase in global temperatures has led to shifts in precipitation patterns, altered the timing and intensity of seasonal rainfall, and intensified evaporation rates, which have all contributed to changes in freshwater availability. Areas that were previously water-rich, such as parts of Africa and the Middle East, are experiencing more frequent and severe droughts, while other regions, such as the Asia-Pacific, have seen increased flooding due to extreme weather events (IPCC, 2021). The shrinking of glaciers and snowpacks, particularly in the Himalayas and the Andes, further reduces the availability of meltwater that millions of people rely on for drinking and irrigation (Immerzeel et al., 2020). The disruption of natural hydrological cycles due to climate change will continue to challenge water management systems and exacerbate regional water scarcity in the coming decades.

2.2.4 Water scarcity and human impact

Water scarcity is not only an issue of availability but also a question of access, equity, and quality. As the global population continues to grow, the demand for freshwater resources intensifies, leading to competition between agricultural, industrial, and domestic users. This competition is often exacerbated by poor water management practices, inadequate infrastructure, and pollution, which limit the availability of clean, safe water. Moreover, many regions face the challenge of ensuring equitable access to water, with marginalized communities, especially in rural areas, having limited access to reliable water sources (United Nations, 2015). Water scarcity is closely linked to socio-economic development, food security, and human health, and its consequences can lead to social and political tensions, economic losses, and the displacement of populations.

While the global availability of freshwater is a well-researched topic, knowledge gaps persist, particularly in the areas of water governance, the impact of climate change on regional water cycles, and the implications of water scarcity for human health and social equity. While considerable literature exists on the technical and environmental aspects of water management, the social dimensions, such as how water scarcity exacerbates inequalities or how water governance systems can be improved, remain less explored (Mason et al., 2016). This book aims to bridge these gaps by offering a comprehensive analysis of global freshwater availability that includes not only the environmental and technical aspects but also the social, political, and economic implications of water scarcity.

2.3 Distribution Patterns of Freshwater Resources

The global distribution of freshwater resources is highly uneven, with some regions of the world experiencing abundant supplies of freshwater, while others are facing severe scarcity. These disparities are shaped by a combination of natural and human-induced factors, including climate, geography, hydrological cycles, population density, and water management practices.

2.3.1 Natural factors: Climate and Geography

Climate and geographical location are the primary natural determinants of freshwater distribution. Regions that are naturally arid or semi-arid, such as the Middle East, North Africa, and parts of Central Asia, face severe freshwater scarcity. The lack of rainfall in these regions limits the availability of surface water in rivers and lakes, which are the most accessible sources of freshwater. In contrast, areas with abundant rainfall, such as tropical rainforests in the Amazon Basin and Southeast Asia, receive large volumes of precipitation, resulting in high freshwater availability.

For instance, the Amazon River basin in South America, one of the world's largest freshwater systems, benefits from the region's equatorial climate, which produces consistent rainfall year-round, replenishing rivers and groundwater supplies (Fearnside, 2005). Similarly, Southeast Asia's monsoon climate leads to annual rainfall patterns that sustain large freshwater reserves, contributing to the abundance of water in countries like Indonesia and Malaysia.

On the other hand, regions in desert climates, such as the Sahara Desert in North Africa, are characterized by extremely low rainfall, often below 50 mm per year. These areas receive insufficient precipitation to replenish freshwater systems, resulting in extreme water scarcity. The Arabian Peninsula, including countries like Saudi Arabia and Yemen, is another example where desert conditions lead to low natural freshwater availability (Mekonnen & Hoekstra, 2016).

2.3.2 Human factors: Population density, urbanization, and water use

While natural factors play a significant role in water availability, human-induced factors such as population density, urbanization, and water consumption patterns also shape the distribution of freshwater resources. Regions with high population densities and urbanization, such as parts of South Asia (e.g., India and Pakistan) and East Asia (e.g., China), often face significant pressure on their freshwater resources. High demand for water for drinking, agriculture, and industry can lead to unsustainable extraction from rivers and groundwater, exacerbating regional scarcity (Sengupta, 2018).

In contrast, countries with lower population densities and better water management systems, such as Canada, Russia, and Brazil, tend to have an abundance of freshwater. These countries have large freshwater reserves due to their vast landscapes and relatively low population pressures, allowing for more equitable distribution of water resources. For example, Canada holds nearly 20% of the world's renewable freshwater supply, with abundant lakes and rivers spread across a sparsely populated country (Postel, 2017).

2.3.3 Water mismanagement and infrastructure issues

Water scarcity in some regions is exacerbated not only by natural conditions but also by poor water management practices. Mismanagement, such as inefficient irrigation systems, over-extraction of groundwater, and pollution of freshwater resources, can lead to a depletion of available water. In many developing countries, inadequate water infrastructure, insufficient investment in water conservation technologies, and lack of effective governance result in water losses and poor distribution. For example, in India, excessive groundwater extraction for agricultural irrigation has led to falling water tables, making freshwater less accessible for both drinking and farming (Bierkens, 2015).

In contrast, countries with well-developed water management systems, such as Singapore and Israel, have been able to overcome natural water scarcity through innovative practices like desalination, water recycling, and efficient irrigation techniques. Israel, for instance, has invested heavily in desalination plants and drip irrigation technologies, allowing it to manage its limited freshwater resources more effectively (Lazarova & Bahri, 2017).

2.3.4 Geopolitical factors: Shared water resources

Many regions face water scarcity due to geopolitical factors, particularly in areas where multiple countries share transboundary rivers or groundwater resources. In regions like the Nile River Basin in Africa, the Tigris-Euphrates Basin in the Middle East, and the Indus River Basin in South Asia, water resources are often a source of conflict due to competing national interests. These shared water resources can lead to tensions over water allocation, particularly as countries struggle to meet the needs of their growing populations and economies. The management of transboundary rivers often requires international cooperation and negotiation, which can be hindered by political instability and historical disputes (Wolf et al., 1999).

2.3.5 Climate change and future water scarcity

Climate change is expected to exacerbate the existing disparities in water availability. Rising global temperatures and changing precipitation patterns are likely to affect water supply and demand in various regions. Areas that are already water-scarce may experience even less rainfall, leading to further depletion of freshwater reserves. For example, climate models predict that the Mediterranean region, including parts of Spain, Italy, and Greece, will become drier, resulting in more frequent and severe droughts (IPCC, 2021). Conversely, some regions may experience increased rainfall, leading to more flooding, but these water surpluses may not be effectively captured or stored for future use, exacerbating water management challenges.

The distribution of freshwater resources across the globe is influenced by a complex interplay of natural, human, and geopolitical factors. While some regions enjoy an abundance of freshwater due to favorable climatic conditions, others face extreme scarcity due to low rainfall, high population pressures, and poor water management practices. Addressing water scarcity will require a multifaceted approach that includes improved water governance, the adoption of sustainable water management practices, and international cooperation to ensure equitable access to water for all.

2.4 Factors Influencing Water Access: Climate, Geography, Population Density, Industrial and Agricultural Demand, and Infrastructure

Water access is determined by a complex array of factors that interplay to either enhance or limit the availability of freshwater. While natural conditions such as climate and geography play a pivotal role, human factors—such as population density, industrial and agricultural demand, and infrastructure—also have significant impacts on water distribution and accessibility. This section explores the key factors influencing water access.

2.4.1 Climate

Climate is one of the most significant natural factors that determine water availability in a region. Precipitation patterns, temperature, and evaporation rates all contribute to how much water is available in a given area. In regions with abundant rainfall, such as tropical climates near the equator, water is often more readily available in rivers, lakes, and groundwater. For example, the Amazon Basin receives heavy rainfall throughout the year, making freshwater resources abundant for the surrounding population (Fearnside, 2005).

Conversely, arid and semi-arid regions, such as the Middle East and North Africa, experience much lower levels of rainfall. These areas face chronic water scarcity, with insufficient natural replenishment of freshwater resources. Climate change is expected to exacerbate these challenges by altering rainfall patterns, increasing temperatures, and raising evaporation rates, leading to even greater water stress in already vulnerable regions (IPCC, 2021).

2.4.2 Geography

Geography, including the location of freshwater sources, topography, and proximity to water bodies, significantly influences water access. Some regions have an abundance of natural freshwater resources, such as large rivers, lakes, and groundwater aquifers. For instance, Canada has one of the world's largest supplies of freshwater, with vast lakes and rivers across its largely uninhabited landscape (Postel, 2017). Similarly, Russia, with its extensive network of rivers and lakes, enjoys ample freshwater resources.

In contrast, landlocked countries or those situated in deserts often lack major rivers or lakes, making it harder to access freshwater. The geography of groundwater aquifers also plays a role, with some areas hosting vast underground water reserves, while others, particularly arid regions, have little to no groundwater sources. For example, the Arabian Peninsula has limited freshwater sources, relying heavily on desalination for its water supply (Mekonnen & Hoekstra, 2016).

2.4.3 Population density

Population density directly affects water demand and access. In densely populated regions, the demand for freshwater for domestic, industrial, and agricultural purposes is higher, often putting significant pressure on local water resources. High population densities in urban areas, particularly in developing countries, can lead to over-extraction of water from rivers and groundwater, causing depletion of local resources. For example, in India, which has a high population density, there is increasing competition for water, with many regions experiencing groundwater depletion due to agricultural and urban demands (Bierkens, 2015).

In contrast, countries with low population densities, such as Canada or Australia, tend to have more available water per capita, even if they face challenges like seasonal variation in rainfall. The ability to manage and distribute water effectively in densely populated regions is a key factor in addressing water scarcity.

2.4.4 Industrial and agricultural demand

Water usage in agriculture and industry is a major contributor to water stress, especially in regions where these sectors are heavily reliant on freshwater. Agriculture accounts for around 70% of global freshwater withdrawals, with water-intensive crops and livestock production requiring large amounts of water (FAO, 2020). In regions such as the American Midwest and India, where irrigation plays a central role in food production, water is often over-extracted from rivers and aquifers to meet agricultural needs.

Industrial demand for water can also be significant, particularly in water-intensive industries such as manufacturing, mining, and energy production. For instance, the water required for cooling power plants and manufacturing processes can place significant pressure on local freshwater resources. As industries expand, particularly in rapidly developing regions, the competition for water between domestic, industrial, and agricultural uses can lead to increased water scarcity.

In regions like the Nile River Basin, where agriculture, industry, and domestic water use are all concentrated around a limited water source, managing the demands of these sectors is critical to maintaining sustainable water access (Wolf et al., 1999).

2.4.5 Infrastructure

The availability of water infrastructure—such as water treatment plants, pipelines, reservoirs, and irrigation systems—plays a crucial role in determining access to freshwater. Even in regions with abundant water resources, poorly managed or outdated infrastructure can prevent equitable distribution and cause water shortages. For example, in many developing countries, inadequate infrastructure can lead to significant water losses due to leakage, contamination, or lack of efficient water storage and distribution systems. In such regions, even if water resources are present, access to clean water can be limited due to infrastructural constraints.

On the other hand, well-developed water infrastructure can help manage water efficiently and ensure equitable distribution, especially in urban areas. For instance, Singapore, despite its limited natural water resources, has developed an advanced water management system that includes desalination, water recycling, and efficient distribution networks to ensure a reliable water supply (Lazarova & Bahri, 2017).

2.4.6 Human governance and water management

Beyond natural and infrastructural factors, effective governance and water management are key to ensuring equitable access to freshwater. Regions with robust water management policies that emphasize conservation, efficient use, and equitable distribution tend to have better access to water resources. For example, Israel's water management practices, which include the use of desalination, drip irrigation, and water recycling, have enabled the country to overcome its natural water scarcity (Lazarova & Bahri, 2017).

However, in regions with poor governance, corruption, or mismanagement, water resources may be allocated ineffectively or unfairly, exacerbating water scarcity. For example, water mismanagement in parts of India has led to over-extraction of groundwater, resulting in a decline in water availability in some areas (Sengupta, 2018).

2.4.7 Examples of influencing factors of water access

A table summarizing exemplary cases with locations, highlighting factors influencing water access such as climate, geography, population density, industrial and agricultural demand, and infrastructure:

This table highlights how different factors influence water access in various regions and provides academic references to support the analysis. Each location is a prime example of how these factors intersect to either exacerbate or alleviate water scarcity or abundance.

This table provides a comprehensive summary of exemplary cases illustrating how various factors influence water access across different regions. The factors include climate, geography, population density, industrial and agricultural demand, and infrastructure, which are key determinants in whether a region experiences water scarcity or abundance. The table not only outlines specific locations but also offers a detailed description of how these factors shape water availability in each case.

Amazon Basin, Brazil: The tropical climate in the Amazon Basin results in significant rainfall, which provides the region with an abundant supply of freshwater. However, the issue of deforestation, as discussed by Fearnside (2005), is a major concern in the region, affecting water cycles and ecosystem sustainability. Despite the natural abundance, the broader environmental context impacts water management.

Table 2.1 Examples of influencing factors of water access

Location	Factor Influencing Water Access	Description	
Amazon Basin, Brazil	**Climate**	The region benefits from a tropical climate with heavy rainfall, resulting in abundant freshwater resources.	Fearnside, P. M. (2005).
Middle East and North Africa	**Climate**	The region faces arid and semi-arid conditions with limited rainfall, leading to significant water scarcity.	IPCC. (2021).
Canada	**Geography**	Canada has vast freshwater resources, including large lakes and rivers, due to its geography and abundant natural water sources.	Postel, S. (2017).
Arabian Peninsula	**Geography**	Located in a desert region, the Arabian Peninsula has limited freshwater resources, relying heavily on desalination.	Mekonnen, M. M., & Hoekstra, A. Y. (2016).
India (Punjab Region)	**Population Density**	With a high population density and intensive agricultural practices, water resources are over-exploited, leading to depletion of groundwater.	Bierkens, M. F. P. (2015).
Singapore	**Infrastructure**	Singapore has developed a highly efficient water management system, including desalination and water recycling, to address water scarcity.	Lazarova, V., & Bahri, A. (2017).
Nile River Basin	**Industrial and Agricultural Demand**	The Nile River supports high agricultural and industrial water demands, which places significant pressure on its water resources.	Wolf, A. T., et al. (1999).
Sub-Saharan Africa	**Population Density & Infrastructure**	Many countries face challenges due to high population growth and inadequate water infrastructure, leading to poor access to clean water.	FAO. (2020).
California, USA	**Industrial and Agricultural Demand**	The state faces competition for water between agriculture, industry, and urban areas, exacerbating water scarcity.	Postel, S. (2017).

Middle East and North Africa: This region is characterized by arid and semi-arid climates, leading to limited natural freshwater resources. As the IPCC (2021) notes, water scarcity is exacerbated by climate change, which alters rainfall patterns, making it harder for many countries in this region to maintain sustainable water supplies.

Canada: With vast natural freshwater resources, including lakes and rivers, Canada is generally fortunate in terms of geography. Postel (2017) highlights that the country's geography provides a significant buffer against water scarcity, but regional disparities exist in access due to varying infrastructure and seasonal distribution of water.

Arabian Peninsula: The Arabian Peninsula faces a significant challenge in accessing freshwater, as it is located in a desert region. With limited natural freshwater sources, countries in this area, such as Saudi Arabia, rely heavily on desalination technology to meet water demands (Mekonnen & Hoekstra, 2016). This reliance on technology highlights the intersection of geography and infrastructure in ensuring water access.

India (Punjab Region): The Punjab region of India faces extreme water scarcity due to high population density and extensive agricultural practices. Groundwater depletion is a growing concern as agricultural demands outstrip the available resources. Bierkens (2015) suggests that over-exploitation of water resources, particularly groundwater, is a key driver of the region's water stress.

Singapore: Despite being a small island nation with limited natural water resources, Singapore has developed an exemplary water management infrastructure. Through desalination, water recycling, and other efficient practices, the country has successfully mitigated its water scarcity challenges (Lazarova & Bahri, 2017). This example highlights how advanced infrastructure can address water challenges even in regions with limited natural resources.

Nile River Basin: The Nile River is a critical water source for several countries, but industrial and agricultural demands place enormous pressure on its waters. The basin supports intensive agricultural production and has significant industrial needs, making water management a critical issue for the countries that depend on it (Wolf et al., 1999).

Sub-Saharan Africa: Many countries in Sub-Saharan Africa experience water scarcity due to high population growth and inadequate infrastructure. FAO (2020) emphasizes that while the region faces growing water demand due to population increase, the lack of investment in infrastructure exacerbates the problem, limiting access to clean water.

California, USA: California faces a complex water scarcity issue, influenced by both industrial and agricultural demand. The competition for water between agriculture, industry, and urban areas leads to tensions and inefficiencies in water distribution. As Postel (2017) discusses, water scarcity in California is aggravated by these competing interests, making it a prime example of how industrial and agricultural demand can strain water resources.

This table illustrates the diversity of factors influencing water access across different geographical regions, with each case providing unique insights into how natural and human-made factors intersect. The academic references provide a solid foundation for understanding these dynamics and highlight the importance of context-specific solutions for managing water resources effectively.

Water access is influenced by various factors, ranging from natural conditions like climate and geography to human activities such as industrial and agricultural demand, population density, and infrastructure. Effective water management, including developing appropriate infrastructure and adopting sustainable water practices, is essential for ensuring equitable access to freshwater in both water-abundant and water-scarce regions. Addressing the growing challenge of water scarcity will require a comprehensive approach that considers both the natural and human-induced factors influencing water availability.

References

Bierkens, M. F. P. (2015). Global hydrology: Trends and progress. *Nature Geoscience*, 8(10), 10–13. https://doi.org/10.1038/ngeo2547

FAO. (2020). *The State of the World's Water Resources for Food and Agriculture*. Food and Agriculture Organization of the United Nations.

Fearnside, P. M. (2005). "Deforestation in the Brazilian Amazon." *Environmental Conservation*, 32(4), 379–387. https://doi.org/10.1017/S0376892905002842

Gleick, P. H. (1993). "Water in crisis: A guide to the world's fresh water resources." *Oxford University Press*.

Gleick, P. H. (2019). *The World's Water: The Biennial Report on Freshwater Resources*. Island Press.

Immerzeel, W. W., et al. (2020). "Glacier mass loss in the Hindu Kush–Karakoram–Himalaya." *Science Advances*, 6(36), eabb3217. https://doi.org/10.1126/sciadv.abb3217

IPCC. (2021). *Climate Change 2021: The Physical Science Basis*. Intergovernmental Panel on Climate Change.

Lazarova, V., & Bahri, A. (2017). Water reuse: From the basics to the new era of integrated resource recovery. *Water International*, 42(7), 911–922. https://doi.org/10.1080/02508060.2017.1348203

Mason, S., et al. (2016). "Water governance: From crises to solutions." *Water Policy*, 18(2), 247–266. https://doi.org/10.2166/wp.2016.171

Mekonnen, M. M., & Hoekstra, A. Y. (2016). Four billion people facing severe water scarcity. *Science Advances*, 2(2), e1500323. https://doi.org/10.1126/sciadv.1500323

Postel, S. L. (2017). *Replenish: The Virtuous Cycle of Water and Prosperity*. Island Press.

Rogers, P., & Hall, A. W. (2003). *Effective Water Governance*. Global Water Partnership.

Scanlon, B. R., Faunt, C. C., Longuevergne, L., Reedy, R. C., Alley, W. M., McGuire, V. L., & McMahon, P. B. (2012). "Groundwater depletion and sustainability of irrigation in the US High Plains and Central Valley." *Proceedings of the National Academy of Sciences*, 109(24), 9320–9325.

Sengupta, R. (2018). "Water Scarcity and Its Impact on India." *The Indian Economic Review*, 53(1), 67–91. https://doi.org/10.1007/s42200-018-0006-4

Shiklomanov, I. A. (1993). "World fresh water resources." In *Water in Crisis: A Guide to the World's Fresh Water Resources* (pp. 13–24). Oxford University Press.

UNESCO. (2020). *World Water Development Report 2020: Water and Climate Change*. United Nations Educational, Scientific and Cultural Organization.

United Nations. (2015). "Water for Sustainable Living." *United Nations Water*.

United Nations. (2018). "The United Nations World Water Development Report 2018: Nature-Based Solutions for Water." *UNESCO*. https://doi.org/10.18356/9789231002747

Wolf, A. T. (2007). "Shared waters: Conflict and cooperation." *Annual Review of Environment and Resources*, 32(1), 241–269.

Wolf, A. T., et al. (1999). "International river basins of the world." *International Journal of Water Resources Development*, 15(4), 405–427. https://doi.org/10.1080/07900629948922

WWF. (2020). *Living Planet Report 2020*. World Wildlife Fund.

3

Extreme Water Scarcity Regions

Water scarcity is a growing global concern, affecting both the availability and accessibility of freshwater resources. Extreme water scarcity regions are those where the demand for water surpasses the available supply, leading to severe challenges in managing water resources for basic human needs, agriculture, and industrial purposes. These regions are primarily located in arid and semi-arid climates, where natural freshwater sources are limited, and human activities further exacerbate the scarcity. This section examines some of the regions facing extreme water scarcity, identifying the factors contributing to these challenges.

3.1 Middle East and North Africa (MENA)

The Middle East and North Africa is one of the most water-scarce regions globally, where water stress is a result of a combination of climate, geography, and over-exploitation of available resources. The region is largely arid or semi-arid, with very limited rainfall and few freshwater sources. As a result, many MENA countries face critical water shortages, with water availability significantly below the global average (Mekonnen & Hoekstra, 2016). For example, countries like Saudi Arabia and the United Arab Emirates rely heavily on desalination plants to meet domestic and industrial water needs. However, desalination is energy-intensive and costly, which poses additional challenges for sustainable water management. Furthermore, the growing population and increasing industrial and agricultural demands are placing further pressure on already limited water resources (El-Fadel et al., 2003).

The IPCC (2021) reports that climate change is likely to intensify water scarcity in the region, with increasing temperatures and altered precipitation patterns leading to more frequent droughts and reduced water availability. Consequently, countries in the MENA region are exploring alternative water

DOI: 10.1201/97818743801061-3

sources, such as treated wastewater, but significant infrastructure barriers and public acceptance of such measures remain.

3.1.1 Reasons for scarcity: Arid climate, high evaporation rates, over-extraction, limited renewable water sources

Water scarcity is a complex phenomenon that arises from both natural and human-induced factors. In many regions of the world, a combination of arid climates, high evaporation rates, over-extraction of water, and the limited availability of renewable water sources contribute significantly to water stress. This section will elaborate on the reasons for water scarcity, providing insight into how and why these factors exacerbate the challenges of managing freshwater resources.

3.1.1.1 Arid climate

Arid climates, which are characterized by low rainfall and high temperatures, are among the primary contributors to water scarcity. In regions with arid or semi-arid climates, such as the Middle East and North Africa (MENA), water availability is naturally limited. These regions receive very little rainfall, with some areas receiving less than 100 millimeters annually, resulting in insufficient natural water replenishment for rivers, lakes, and groundwater sources (IPCC, 2021).

The lack of precipitation in arid climates means that the natural hydrological cycle is disrupted, leading to a constant state of water scarcity. This is particularly problematic in areas where agriculture and human settlements are concentrated, as these regions rely heavily on water for irrigation and daily consumption. For instance, countries like Saudi Arabia and Egypt have to manage water resources carefully due to the absence of significant rainfall and limited natural water sources, making them highly vulnerable to fluctuations in water supply (Mekonnen & Hoekstra, 2016).

3.1.1.2 High evaporation rates

High evaporation rates further exacerbate water scarcity, particularly in regions with high temperatures and dry conditions. Evaporation refers to the process by which water from the earth's surface (including lakes, rivers, and reservoirs) changes into vapor and rises into the atmosphere. In hot and dry climates, the evaporation rate can be so high that water bodies lose significant amounts of water even before it can be used for human or agricultural purposes.

Evaporation rates are influenced by temperature, wind speed, and humidity, and in arid regions, these factors are often intensified. For example, in regions like the Mediterranean Basin or parts of Africa, the combination of high temperatures and low humidity leads to substantial evaporation losses, significantly reducing the usable volume of freshwater (Gao & Shi, 2018). In some areas, such as the Dead Sea in the Jordan River Basin, evaporation has led to the rapid shrinking of water bodies, aggravating water scarcity issues and impacting both ecosystems and local communities (Gertner, 2014).

3.1.1.3 Over-extraction of water

One of the leading causes of water scarcity in both arid and non-arid regions is the over-extraction of water from rivers, groundwater, and other freshwater sources. Over-extraction occurs when water is withdrawn faster than it can be replenished through natural processes like rainfall or river flow. Excessive groundwater extraction is a particularly acute issue in many parts of the world, including India, the United States, and parts of China.

In regions like India's Punjab and the southwestern United States, intensive agricultural practices, combined with growing urban populations, have led to the depletion of groundwater resources (Bierkens, 2015). In Punjab, for example, irrigation systems that rely heavily on groundwater extraction have caused water tables to drop significantly, and many areas now face the risk of complete groundwater depletion. The over-extraction of groundwater leads to a negative water balance, where the water withdrawn exceeds the amount naturally replenished by precipitation or natural recharge, further deepening the scarcity (Konikow, 2013).

Additionally, in some regions, over-extraction for industrial use and the growing demand for drinking water in urban areas also place immense pressure on water resources. This increases competition for water, often leading to conflicts and inequitable distribution.

3.1.1.4 Limited renewable water sources

The availability of renewable water sources is another critical factor in determining water scarcity. Renewable water sources are those that are replenished naturally through rainfall, snowmelt, or groundwater recharge. However, many regions suffer from limited or unevenly distributed renewable water.

Regions such as parts of Sub-Saharan Africa, the Arabian Peninsula, and Central Asia experience significant limitations in renewable water resources. These areas either have very little natural water recharge or rely on distant water bodies or seasonal rainfall that is highly unpredictable. According to the Food and Agriculture Organization (FAO, 2020), many African countries, for instance, experience both low rainfall and high variability in their freshwater supplies, resulting in the inability to sustain large-scale agriculture or meet basic water needs.

In many cases, countries that depend on transboundary water resources (such as the Nile or the Euphrates rivers) face geopolitical challenges that complicate the management of shared water supplies, further reducing the availability of renewable water. As countries and regions compete for access to these resources, it becomes increasingly difficult to ensure fair and sustainable water distribution (Wolf, 1999). Moreover, climate change is expected to exacerbate this situation by altering precipitation patterns and making renewable water resources more unpredictable and less reliable (IPCC, 2021).

Water scarcity is a pressing global challenge, with several interconnected factors contributing to its severity. Arid climates and high evaporation rates naturally limit the availability of freshwater, while over-extraction of water exacerbates the depletion of these resources. In regions with limited renewable water sources, the growing demand for water driven by agricultural, industrial, and domestic needs further deepens the crisis. Addressing water scarcity requires a multifaceted approach, focusing not only on improving water management practices but also on mitigating the impacts of climate change and promoting the equitable distribution of water resources.

3.1.2 Comparative overview of water scarcity in MENA region

The Middle East and North Africa (MENA) region faces some of the most severe water scarcity challenges in the world. Due to its arid and semi-arid climate, compounded by political, economic, and demographic pressures, the MENA region exhibits a variety of water-related challenges across its countries. This comparative overview will examine the causes and consequences of water scarcity in the region, highlighting the factors that contribute to water stress in individual MENA countries, including climate, population growth, political instability, and over-extraction of water resources.

3.1.2.1. General water scarcity context in MENA

MENA is home to approximately 6% of the global population but only holds about 1.5% of the world's renewable freshwater resources (Mekonnen & Hoekstra, 2016). The region is characterized by its arid climate, with much of its landmass experiencing low and irregular rainfall. Water scarcity in MENA countries is primarily driven by climatic conditions, compounded by high population growth, intensive agricultural demands, and over-extraction of groundwater. Furthermore, political instability and regional conflicts often exacerbate the management of shared water resources, especially in trans-boundary river basins like the Nile, Tigris-Euphrates, and Jordan.

3.1.2.2 Country-specific water scarcity challenges

3.1.2.2.1 Egypt

Egypt, situated in the northeastern part of the MENA region, is heavily dependent on the Nile River for 95% of its water needs (Elhadi et al., 2020). Despite being one of the most populous countries in the region, Egypt faces significant challenges in securing sustainable water sources due to the trans-boundary nature of the Nile, the rapid population growth, and the increasing agricultural demand. The country's water supply per capita is among the lowest globally, hovering around 600 cubic meters annually, far below the international threshold for water scarcity, which is set at 1,000 cubic meters per person (Mekonnen & Hoekstra, 2016).

In addition to political disputes over Nile water sharing with upstream countries, climate change impacts such as reduced Nile flow and increased evaporation rates further threaten water availability (Elhadi et al., 2020). Egypt is increasingly turning to alternative sources such as desalinated water and wastewater reuse to address this shortage, but these measures are costly and energy-intensive.

3.1.2.2.2 Saudi Arabia

Saudi Arabia is one of the most water-scarce countries in the world, with virtually no renewable freshwater resources. The country's water supply largely depends on desalination plants and non-renewable groundwater. The country relies on groundwater extraction from deep aquifers, which are being over-exploited for irrigation and urban consumption, which significantly depletes these resources (Fahmy, 2019).

The arid desert climate, characterized by extremely low rainfall, makes it difficult for natural replenishment of water sources, while high evaporation rates from reservoirs further exacerbate the situation. Saudi Arabia has been investing heavily in desalination, producing around 18% of the world's desalinated water, but this is a costly and energy-intensive solution. As a result, water management policies have been central to the country's efforts to address its water crisis, including a shift toward more efficient water usage and large-scale projects for wastewater reuse (Fahmy, 2019).

3.1.2.2.3 Jordan

Jordan is another MENA country experiencing extreme water scarcity. With water resources per capita at less than 100 cubic meters per year, the country is one of the most water-stressed in the world (Haddad et al., 2017). The main sources of water are the Jordan River, the Disi Aquifer, and the Yarmouk River, which are subject to political tensions with neighboring countries. Climate change projections predict that the country will face further reductions in precipitation, increasing its vulnerability to water shortages (Haddad et al., 2017).

In Jordan, urbanization and refugee influxes have led to increased water demand. The influx of refugees from Syria has placed additional pressure on an already strained water supply system. The government has implemented water-saving measures, including the use of desalinated water and recycling, but infrastructure challenges and lack of sufficient funding hinder the sustainability of these solutions (Haddad et al., 2017).

3.1.2.2.4 Iraq

Iraq faces significant challenges regarding water scarcity due to its reliance on the Tigris and Euphrates Rivers, both of which flow through Turkey and Syria, making Iraq dependent on transboundary water sources. Climate change, reduced river flow, and upstream dams have led to declining water availability, placing pressure on Iraq's agricultural and urban water demands (Al-Ansari, 2020).

The over-extraction of groundwater and poor water management practices further exacerbate the situation, leading to the contamination of groundwater resources and the inability of many Iraqis to access clean water. With high temperatures and limited rainfall, Iraq's water crisis is expected to worsen unless regional cooperation on water-sharing agreements is achieved (Al-Ansari, 2020).

3.1.2.2.5 United Arab Emirates (UAE)

The UAE, like Saudi Arabia, is heavily dependent on desalinated water for its freshwater supply. With its arid climate and high evaporation rates, the UAE faces major challenges in securing sustainable water sources. Overreliance on desalination places significant strain on the country's energy resources and poses environmental concerns, particularly the disposal of brine waste into the sea, which harms marine ecosystems (El Maayar et al., 2018).

Additionally, population growth and economic development in urban areas increase the demand for water, with the industrial sector also being a major consumer. As such, the UAE is investing in water recycling and efficient water management practices, but its reliance on desalination means that water security remains a persistent challenge (El Maayar et al., 2018).

The MENA region as a whole is a hotspot of water scarcity, driven by a combination of climatic factors, over-extraction of water, political tensions, and rapid population growth. While countries like Saudi Arabia and the UAE rely heavily on desalination, others, such as Egypt and Iraq, face challenges from transboundary water disputes and climate change. Jordan and other countries are also struggling with the intensifying effects of water scarcity due to increasing urbanization and refugee influxes. Therefore, addressing the region's water challenges requires both innovative water management strategies and regional cooperation.

3.1.3 Impact of water scarcity in the MENA region: Reliance on desalination, conflict over water resources, and severe agricultural limitations

The Middle East and North Africa (MENA) region is widely regarded as one of the most water-scarce regions in the world. The scarcity of freshwater resources significantly impacts the social, economic, and political dynamics of the region. This article examines three critical consequences of water scarcity in the MENA region: (1) reliance on desalination, (2) conflict over shared water resources, and (3) severe agricultural limitations.

3.1.3.1 Reliance on desalination

Desalination has become an essential method of providing freshwater in water-scarce countries in the MENA region. With most of the region's nations situated in arid zones, desalination—especially through reverse osmosis technology—has enabled countries like Saudi Arabia, the United Arab Emirates (UAE), Kuwait, and Qatar to secure freshwater supplies. MENA countries collectively produce about 60% of the world's desalinated water (Fahmy, 2019).

3.1.3.2 Impacts of desalination

Desalination removes salt and impurities from seawater and offers a crucial solution for water-scarce regions. However, its impacts are multifaceted. Positively, it provides a reliable freshwater source, particularly in arid areas facing severe droughts. Yet, it also poses environmental challenges, such as high energy consumption, greenhouse gas emissions, and brine discharge, which can harm marine ecosystems. The concentrated brine byproduct, often laden with chemicals, can increase ocean salinity, affecting aquatic life. Moreover, the high costs associated with desalination technology limit its accessibility, especially for developing countries. Sustainable approaches are needed to mitigate its ecological footprint while harnessing its benefits (Lattemann & Höpner, 2008).

3.1.3.3 Energy consumption

Desalination is highly energy-intensive. The UAE and Saudi Arabia, in particular, rely on fossil fuels to power their desalination plants, which contributes to both environmental degradation and rising carbon emissions (El Maayar et al., 2018). This energy demand is problematic in a region where energy consumption is already high, and it places additional pressure on fossil fuel resources.

Desalination is a critical technology for providing freshwater in water-scarce regions, especially in the Middle East, where arid conditions and limited natural water resources make it a vital component of water security. However, desalination is an energy-intensive process, requiring substantial amounts of energy to extract salt and impurities from seawater. The United Arab Emirates (UAE) and Saudi Arabia are among the leading countries in desalination capacity, but they predominantly rely on fossil fuels to power their desalination plants. This dependency raises concerns regarding environmental sustainability, carbon emissions, and the long-term availability of fossil fuel resources (El Maayar et al., 2018).

3.1.3.4 High energy demand and environmental impact

The energy-intensive nature of desalination is a significant concern for the UAE and Saudi Arabia, where fossil fuel-powered thermal desalination technologies like Multi-Stage Flash (MSF) and Multi-Effect Distillation (MED) are commonly used. These processes consume vast amounts of energy, often derived from natural gas and oil, contributing to high levels of carbon dioxide (CO_2) emissions. For instance, it is estimated that the energy required for thermal desalination can be up to ten times higher than conventional water treatment methods, resulting in a substantial carbon footprint (Soto, 2019). In the UAE, desalination accounts for up to 20% of total domestic energy consumption, with Saudi Arabia similarly dependent on fossil fuels for water production (El Emam & Dincer, 2018). This high energy demand not only increases greenhouse gas emissions but also exacerbates the region's environmental challenges, including air pollution and global warming.

3.1.3.5 Economic and resource implications

The reliance on fossil fuels for desalination in the UAE and Saudi Arabia also has significant economic implications. As global energy demand rises and fossil fuel reserves diminish, the cost of producing desalinated water is expected to increase. This places additional pressure on national economies that are already heavily reliant on oil and gas exports for revenue. The International Energy Agency (IEA, 2019) notes that, in the long term, the unsustainable use of fossil fuels for desalination could threaten energy security, particularly if alternative energy sources are not integrated into the desalination sector. The substantial energy requirements also divert valuable fossil fuel resources from export markets, potentially impacting economic stability in the region.

3.1.3.6 Potential for renewable energy integration

To address the environmental and economic challenges posed by the energy-intensive nature of desalination, there is growing interest in integrating renewable energy sources, such as solar and wind, into desalination processes. The Gulf region, with its abundant solar resources, is particularly well-positioned to leverage solar power for desalination. Emerging technologies like solar-powered reverse osmosis (RO) are gaining attention as more energy-efficient and sustainable alternatives to conventional thermal desalination (Al-Karaghouli & Kazmerski, 2013). Saudi Arabia, for example, has launched initiatives to develop solar-powered desalination plants, aiming to reduce its carbon footprint and decrease its reliance on fossil fuels (Alkaisi et al., 2017). Transitioning to renewable energy-powered desalination could not only alleviate the environmental impacts but also ensure a more sustainable water supply for the future.

While desalination remains a crucial solution for addressing water scarcity in the UAE and Saudi Arabia, its high energy consumption, driven by fossil fuel dependency, poses significant environmental and economic challenges. The ongoing reliance on fossil fuels for desalination contributes to rising carbon emissions and environmental degradation, underscoring the need for a shift towards more sustainable practices. Integrating renewable energy into desalination processes presents a viable pathway to reduce environmental impacts and ensure long-term water security. Continued investment in energy-efficient desalination technologies and adopting renewable energy sources will be essential for mitigating the adverse effects associated with the current desalination practices in the region.

3.1.4 Environmental concerns of desalination: Impact of brine discharge on marine ecosystems

The brine byproduct produced by desalination plants is often dumped back into the sea, damaging marine ecosystems (Elhadi et al., 2020). This affects biodiversity, as high salinity levels can harm marine life and disrupt the food chain.

Desalination is a critical solution for water-scarce regions, but it comes with significant environmental challenges, particularly related to the disposal of brine byproducts. Desalination plants generate concentrated brine—a highly saline wastewater mixed with chemicals like anti-scaling agents and chlorine—during the process of converting seawater into potable water. This brine is often discharged directly back into the sea, which can have severe consequences for marine ecosystems (Elhadi et al., 2020).

3.1.4.1 Effects on marine biodiversity

The direct discharge of brine into coastal waters can elevate local salinity levels, creating a hyper-saline environment that many marine organisms cannot tolerate. High salinity can disrupt the osmoregulatory processes of aquatic species, leading to physiological stress, reduced growth rates, or even mortality. The increased salinity can also alter the composition of local marine communities, threatening sensitive species like coral reefs, seagrasses, and mangroves that play crucial roles in maintaining marine biodiversity (Lattemann & Höpner, 2008). Additionally, the discharge of hot brine, which can be warmer than the surrounding seawater, can further exacerbate these negative impacts by creating thermal pollution, leading to habitat degradation.

3.1.4.2 Disruption of the food chain

Brine discharge affects individual species and can disrupt entire food chains in coastal ecosystems. Marine plankton, which forms the base of the food web, are particularly vulnerable to changes in salinity and temperature. A decline in plankton populations can have cascading effects, impacting fish stocks and higher trophic levels, including commercially important species and marine mammals (Jones et al., 2019). Thus, the environmental footprint of desalination extends beyond immediate coastal areas, potentially affecting the livelihoods of fishing communities and the overall health of marine environments.

3.1.4.3 Mitigation strategies

There is a growing emphasis on developing sustainable desalination practices to mitigate the environmental impacts of brine discharge. Some strategies include diluting brine with seawater before discharge, using diffusers to reduce the concentration of pollutants, or exploring alternative disposal methods like deep-well injection. Additionally, there is increasing research into using renewable energy-powered desalination technologies and brine management solutions, such as converting brine into valuable byproducts like minerals and salts (Voutchkov, 2018). By adopting such practices, it is possible to reduce the ecological footprint of desalination and protect vulnerable marine ecosystems.

3.1.4.4 Economic costs

Desalinated water is significantly more expensive than natural freshwater sources, making it an expensive resource for governments and consumers alike. This economic burden is particularly challenging for countries facing high population growth and economic pressures.

3.1.4.5 Economic costs of desalination: A burden for governments and consumers

Desalination offers a vital solution for water-scarce regions by providing a reliable source of potable water, but it comes at a high economic cost. The production of desalinated water is significantly more expensive than sourcing freshwater from natural sources such as rivers, lakes, or aquifers. This increased cost is largely due to the high energy requirements, complex infrastructure, and maintenance needed for desalination plants. As a result, desalinated water can be up to three times more expensive than conventional freshwater, placing a financial strain on both governments and consumers (Schneider et al., 2020).

3.1.4.6 Financial implications for governments

For many countries, especially those in the Middle East and North Africa (MENA) region, where desalination is a primary water source, the economic burden is substantial. Governments often subsidize desalinated water to make it affordable for the population, which can lead to significant financial outlays. For example, Saudi Arabia and the United Arab Emirates invest billions of dollars annually in desalination infrastructure to meet their domestic water needs (Nabi et al., 2019). However, as demand for water increases due to rapid population growth and urbanization, these costs are expected to escalate, putting additional pressure on national budgets and diverting resources from other critical sectors like healthcare and education.

3.1.4.7 Challenges for consumers and economic development

The high costs of desalination are often passed on to consumers, resulting in higher water tariffs. In regions where desalinated water constitutes a large portion of the water supply, households and businesses face increased utility bills, which can be a burden, especially for low-income families. Additionally, the higher cost of desalinated water can impact agricultural and industrial sectors that rely on large volumes of water, potentially reducing competitiveness and economic growth (Mujtaba & Shahzad, 2020). This is particularly challenging for countries with limited economic resources, where the high cost of desalination can exacerbate socio-economic inequalities and limit access to affordable water.

3.1.4.8 Cost mitigation strategies

Several strategies are being explored to address the economic challenges associated with desalination. These include investing in energy-efficient technologies, such as reverse osmosis (RO), which requires less energy compared to traditional thermal desalination methods like Multi-Stage Flash (MSF). Additionally, there is growing interest in integrating renewable energy sources, such as solar and wind, to power desalination plants, which can reduce operational costs and mitigate the environmental impact of fossil fuel consumption (El-Emam & Dincer, 2018). The use of innovative technologies and renewable energy integration could help lower the cost of desalinated water, making it a more viable option for water-scarce countries.

While desalination remains a crucial solution for addressing water scarcity, its high economic costs present significant challenges for governments and consumers. The financial burden of desalination infrastructure and operation can strain national budgets and increase water prices, impacting economic development and social equity. Exploring cost-efficient technologies and renewable energy integration is essential for making desalination a more sustainable and affordable solution in the future.

3.1.5 Conflict over water resources

Political conflicts over shared water resources further complicate water scarcity in MENA countries. Many countries in the region rely on transboundary rivers and groundwater aquifers shared with neighboring nations. As the region's water resources become increasingly stressed, tensions over access and control of these resources have escalated.

3.1.5.1 Key conflict areas

3.1.5.1.1 The Nile River

Egypt depends heavily on the Nile River for almost all its water supply. However, upstream countries, particularly Ethiopia, are constructing large-scale hydropower projects, such as the Grand Ethiopian Renaissance Dam (GERD), which could significantly reduce water flow into Egypt. This has caused tensions, as Egypt fears its share of the Nile water will be reduced, exacerbating its water scarcity.

3.1.5.1.2 The Nile River: Egypt's water lifeline and the impact of upstream developments

The Nile River is Egypt's primary source of freshwater, accounting for approximately 97% of the country's water supply. With an arid climate and limited rainfall, Egypt is heavily dependent on the Nile for drinking water, agriculture, and industry (Swain, 2011). However, the construction of large-scale hydropower projects by upstream countries, particularly Ethiopia's Grand Ethiopian Renaissance Dam (GERD), poses a significant threat to Egypt's water security. The GERD, located on the Blue Nile, which contributes about 85% of the Nile's total flow, is expected to be Africa's largest hydroelectric power plant once fully operational.

3.1.5.1.3 Potential reduction in water flow

Ethiopia's construction of the GERD has been a point of contention, especially with Egypt, which fears that the dam will significantly reduce the amount of water flowing downstream. The filling and operation of the GERD could potentially decrease Egypt's annual water supply, impacting agriculture, which consumes over 80% of the country's water resources (Cascão & Nicol, 2016). A reduction in water availability would exacerbate Egypt's already critical water scarcity, where the country is nearing the United Nations' threshold for absolute water scarcity of 500 cubic meters per person per year (El-Nashar & Elyamany, 2018). The prospect of diminished water flow raises concerns over food security, economic stability, and social welfare, especially for the millions of Egyptians who depend on the Nile for their livelihoods.

3.1.5.1.4 Regional tensions and diplomatic efforts

The Nile Basin has historically been a region of cooperation, but the construction of the GERD has strained relations between Ethiopia, Egypt, and Sudan. Egypt has expressed concerns that Ethiopia's unilateral actions could set a precedent for upstream countries to pursue water projects without consulting downstream nations, thereby threatening its historical water rights established under colonial-era agreements (Cascão, 2009). Diplomatic negotiations have been ongoing for over a decade, with involvement from international mediators like the African Union, but a binding agreement on the dam's operation and water allocation has yet to be reached. The lack of consensus highlights the complexities of transboundary water management and the potential for conflict over shared water resources (Tawfik, 2016).

3.1.5.1.5 Mitigating water scarcity

Egypt has been exploring alternative strategies to enhance its water security to mitigate the potential impact of reduced Nile flows. These measures include expanding the use of desalination, improving irrigation efficiency, and implementing water conservation programs. Egypt has also accelerated its investments in wastewater treatment to maximize the reuse of water (Abdel-Shafy & Mansour, 2013). Despite these efforts, the long-term sustainability of Egypt's water resources remains uncertain, particularly if upstream developments on the Nile continue without regional cooperation.

The Nile River remains Egypt's most vital water source, but the construction of the GERD by Ethiopia represents a significant challenge to Egypt's water security. As upstream countries seek to harness the Nile for their own development, it underscores the need for cooperative management of this critical transboundary resource. Without a collaborative framework, the region risks escalating tensions that could threaten not only water security but also regional stability.

3.1.5.2 The Tigris and Euphrates Rivers

Iraq, Syria, and Turkey share the Tigris and Euphrates River basins, but Turkey has built a series of dams, such as the Ataturk Dam, which has reduced water flow to Syria and Iraq. This has led to disputes and calls for more equitable water-sharing agreements.

3.1.5.2.1 The Tigris and Euphrates Rivers: A source of regional tension over water resources

The Tigris and Euphrates rivers are critical freshwater sources for Iraq, Syria, and Turkey, providing water for agriculture, drinking, and industrial uses. These rivers originate in Turkey and flow through Syria and Iraq before emptying into the Persian Gulf, making them a vital shared resource in an otherwise arid region (Kibaroglu, 2019). However, Turkey's extensive dam-building projects, particularly the Southeastern Anatolia Project (GAP), which includes the construction of major dams like the Ataturk Dam, have significantly reduced the flow of water downstream to Syria and Iraq. This has intensified regional disputes over water rights and raised calls for more equitable water-sharing agreements.

3.1.5.3 Impact of Turkey's dams on downstream water availability

Turkey's GAP initiative, which encompasses over 22 dams and 19 hydroelectric power plants, was designed to boost economic development in its southeastern region by providing irrigation and generating electricity (Altinbilek, 2004). The Ataturk Dam, one of the largest in the world, plays a central role in this project. However, the reduced flow of water from the Tigris and Euphrates rivers has had severe consequences for downstream countries. Iraq, for example, has seen a drastic decrease in its water levels, which has led to significant impacts on agriculture, drinking water supplies, and the overall livelihood of its population (Al-Ansari, 2013). Similarly, Syria has faced challenges in sustaining its agricultural sector, which relies heavily on water from these rivers.

3.1.5.3.1 Escalating regional disputes

The unilateral construction of dams by Turkey has exacerbated tensions with Iraq and Syria, both of which accuse Turkey of restricting their access to vital water resources. These tensions have sparked calls for comprehensive water-sharing agreements that take into account the needs of all riparian countries. However, negotiations have been challenging due to the absence of binding international treaties governing the allocation of transboundary river waters in the region (Allan, 2001). The lack of a cooperative framework has led to accusations of water hoarding and unfair practices, further complicating diplomatic relations in a politically volatile region.

3.1.5.3.2 Environmental and social implications

The reduction in water flow from the Tigris and Euphrates has affected agriculture and had broader environmental and social consequences. Reduced river flow has led to the degradation of wetlands, loss of biodiversity, and increased salinity in the soil, which threatens food security in Iraq and Syria (UN-ESCWA & BGR, 2013). Moreover, water scarcity has contributed to the displacement of communities and exacerbated existing social tensions, particularly in Iraq, where competition for dwindling water resources has fueled local conflicts.

3.1.5.3.3 The need for cooperative water management

Addressing the water crisis in the Tigris-Euphrates basin requires a collaborative approach to transboundary water management. Experts have long advocated for the establishment of a comprehensive water-sharing agreement that ensures equitable distribution of water resources among Turkey, Syria, and Iraq (Kibaroglu & Scheumann, 2013). Such an agreement would not only mitigate the risk of conflict but also promote sustainable water use and regional stability. International organizations, including the United Nations, have called for renewed diplomatic efforts to foster cooperation over shared water resources, emphasizing that water scarcity in the Middle East is a critical issue that demands urgent attention.

The Tigris and Euphrates rivers are essential lifelines for Iraq, Syria, and Turkey, yet the region faces significant challenges due to upstream dam constructions. Turkey's ambitious dam projects have reduced water availability for its neighbors, leading to disputes over water rights and calls for equitable sharing agreements. Without a binding framework for transboundary water management, these disputes are likely to persist, threatening the stability and sustainability of the region.

3.1.5.4 Jordan River

The Jordan River is shared between Israel, Jordan, Syria, and Palestine. Water scarcity in this basin has contributed to conflicts, with Israel historically controlling most of the water flow, leading to disputes over rights and usage between the countries.

3.1.5.4.1 The Jordan River: Shared water resources and regional conflicts

The Jordan River is a crucial water source for Israel, Jordan, Syria, and Palestine, serving as a lifeline in one of the world's most water-scarce regions. The river, which stretches for approximately 251 kilometers, plays a vital role in providing freshwater for domestic, agricultural, and industrial use. However, water scarcity and unequal distribution have exacerbated tensions among these riparian states, turning the Jordan River basin into a hotspot for conflicts over water rights and usage (Wolf, 1996). Historical control over the river's flow, particularly by Israel, has led to disputes with neighboring countries, each vying for a fair share of this limited resource.

3.1.5.4.2 Unequal distribution and control of water resources

Israel has historically maintained control over a significant portion of the Jordan River's water flow, primarily through the National Water Carrier project, which diverts water from the Sea of Galilee (Lake Tiberias) to supply various parts of the country (Zeitoun, 2008). This diversion has significantly reduced the amount of water available downstream for Jordan and Palestine, contributing to a chronic state of water scarcity in these regions. For example, Jordan, one of the most water-poor countries in the world, relies heavily on the Jordan River to meet its water needs. However, the reduced flow has led to overexploitation of groundwater resources, further aggravating the country's water crisis (Al-Sharif, 2019).

The Palestinian territories, particularly the West Bank, also face severe water shortages due to restrictions on access to water resources controlled by Israel. Palestinians in the West Bank have limited control over their water infrastructure and must purchase water from Israel at high costs, leading to significant disparities in water availability and quality (Selby, 2013). This situation has fueled grievances and added another layer of complexity to the Israeli-Palestinian conflict.

3.1.5.4.3 Environmental degradation and overuse

The overuse and mismanagement of the Jordan River have led to severe environmental degradation. Over the past decades, the river's flow has decreased by over 90%, primarily due to upstream water diversions by Israel, Syria, and Jordan for agricultural and domestic use (Gafny et al., 2010). This over-extraction has resulted in a dramatic drop in water levels, transforming the once-thriving ecosystem into a depleted and polluted watercourse. The lower Jordan River is now heavily contaminated with untreated sewage, agricultural runoff, and saline water, which has further degraded water quality and threatened biodiversity in the region (Kliot, 2005).

3.1.5.4.4 Efforts toward cooperation and conflict resolution

Despite the challenges, there have been efforts to foster cooperation over the shared waters of the Jordan River. The 1994 Israel-Jordan Peace Treaty included provisions for water-sharing, allowing Jordan to receive a specified amount of water annually from Israel (Phillips et al., 2007). This agreement, while a step forward, has not fully resolved the broader issue of water allocation among all riparian states, particularly Syria and Palestine, which remain outside of formal water-sharing agreements.

Various non-governmental organizations and international bodies have advocated for collaborative water management solutions to address the scarcity issues in the Jordan River basin. Initiatives like the "Good Water Neighbors" project aim to promote cooperation among communities across borders by highlighting the shared benefits of sustainable water use and joint environmental stewardship (Jägerskog & Zeitoun, 2009). However, the success of such initiatives is often hindered by the broader political and security concerns that dominate the region.

The Jordan River serves as both a vital resource and a source of conflict for Israel, Jordan, Syria, and Palestine. Unequal distribution historical control by Israel, and environmental degradation have exacerbated tensions among these countries, making water scarcity a critical issue in the region.

While there have been efforts to promote cooperation, the challenges of achieving equitable water-sharing remain tied to the complex political landscape of the Middle East. Sustainable management of the Jordan River will require not only technical solutions but also diplomatic efforts to address the underlying issues of equity and access.

3.1.5.5 Impacts of water conflicts

Water disputes have strained diplomatic relations, leading to challenges in regional cooperation on other critical issues such as trade, security, and infrastructure development (Al-Ansari, 2020).

3.1.5.5.1 Diplomatic strains: The impact of water disputes on regional cooperation

Water disputes in the Middle East, particularly over shared river basins such as the Jordan, Nile, and Tigris-Euphrates, have exacerbated water scarcity and created significant diplomatic strains. These tensions often transcend water issues, influencing broader regional cooperation on critical matters such as trade, security, and infrastructure development. The control and allocation of water resources have become deeply entangled with political and territorial disputes, complicating efforts to achieve regional peace and stability. As countries vie for control over limited water supplies, these conflicts have far-reaching consequences, affecting not only relations between nations directly involved in the disputes but also their ability to collaborate on shared challenges.

3.1.5.5.2 Water as a source of diplomatic tension

In regions like the Middle East, where water is a scarce and highly contested resource, water disputes are frequently linked to national security concerns. For example, the construction of large-scale dams, like Ethiopia's Grand Ethiopian Renaissance Dam (GERD) on the Nile River, has sparked tensions with downstream countries such as Egypt and Sudan, which fear that the dam will significantly reduce their share of the Nile's water. This has led to protracted diplomatic negotiations, with little resolution in sight, contributing to a broader sense of insecurity in the region (Al-Ansari, 2020). Similarly, the ongoing water-related conflicts between Israel, Jordan, and Palestine over the Jordan River have strained not only bilateral relations but have also hampered efforts to engage in broader regional cooperation.

The political and economic ramifications of water disputes are significant. When water becomes a bargaining chip or a point of contention, it becomes difficult for neighboring countries to collaborate on other important issues, such as trade agreements, security cooperation, or shared infrastructure projects. Water access and control are often seen as symbols of sovereignty, leading to resistance against negotiated settlements that could appear to compromise national interests. These strained diplomatic relations hinder the establishment of trust among neighboring states, which is necessary for broader cooperation in various other sectors.

3.1.5.5.3 Water disputes and regional security

In addition to hindering cooperation on economic and trade issues, water disputes can escalate tensions and contribute to broader security concerns. The limited availability of freshwater resources in the Middle East, combined with the politicization of water management, has made it a security issue in many countries. For instance, Turkey's construction of large dams on the Tigris and Euphrates rivers has led to concerns in Iraq and Syria regarding water shortages and the potential for conflict. Iraq has even considered military options to secure its water supply in the face of Turkey's control over the flow of these rivers (Sabbagh, 2019).

Water-related conflicts also play into larger geopolitical rivalries in the region. As countries increasingly view water as a strategic resource, they are less likely to engage in regional cooperation on water management, which could potentially lead to further instability. Diplomatic efforts to resolve these issues are often undermined by the broader political and security challenges that accompany the water disputes.

3.1.5.5.4 The role of diplomacy in water conflict resolution

Despite the challenges posed by water disputes, there have been instances where diplomacy has played a critical role in alleviating tensions and facilitating cooperation. One notable example is the 1994 Israel-Jordan peace treaty, which included a water-sharing agreement between the two countries. This agreement allowed Jordan to receive a guaranteed amount of water from Israel, which helped to improve bilateral relations and reduce tensions over water access (Phillips et al., 2007). While the treaty was a significant achievement, the complexities of water-sharing arrangements in the region remain unresolved, as issues persist with Palestine and Syria, both of which are not party to formal agreements.

The importance of diplomatic engagement cannot be overstated. Negotiating and implementing equitable water-sharing arrangements can build trust, promote cooperation on other regional issues, and help to address the broader challenges of water scarcity and environmental degradation. However, such efforts require political will, trust-building measures, and an understanding of the interconnectedness of water with broader geopolitical and security concerns.

Water disputes in the Middle East have had profound diplomatic consequences, impeding cooperation on key regional issues such as trade, security, and infrastructure development. The strategic importance of water, compounded by the high levels of scarcity, has made it a contentious resource, often tied to national sovereignty and security. While diplomatic efforts have had some success in resolving water-related conflicts, the broader political dynamics of the region make it challenging to foster long-term, cooperative solutions. The need for fair and equitable water-sharing agreements, underpinned by trust and regional cooperation, remains critical for reducing tensions and ensuring peace and stability in the region.

3.1.5.6 Increased tensions and security risks

In areas where water scarcity and conflict overlap, the risk of political instability and armed conflict increases. Control over water resources is often linked to national security concerns.

3.1.5.6.1 Increased tensions and security risks: The link between water scarcity and conflict

Water scarcity, particularly in regions where resources are already limited, often intersects with political instability, contributing to heightened tensions and increased security risks. In these areas, access to water becomes not just an economic issue but also a matter of national security, as states compete for control over vital water resources. As demand for freshwater rises, and supplies dwindle due to factors like population growth, climate change, and infrastructure challenges, disputes over water can escalate, potentially leading to violent conflict.

3.1.5.6.2 Water as a strategic resource

Water is an essential resource for human survival, economic development, and national security, and when it becomes scarce, it can trigger significant geopolitical tensions. Countries that share transboundary water sources, such as rivers and aquifers, are particularly vulnerable to conflicts. For instance, the Nile River basin, shared by Egypt, Sudan, and Ethiopia, has long been a source of tension, particularly over the construction of the Grand Ethiopian Renaissance Dam (GERD). Ethiopia's plans to build and fill the dam have raised concerns in Egypt and Sudan, which depend heavily on the Nile for their water supply. Egypt, in particular, views any reduction in its share of the

Nile's water as a threat to its national security, leading to diplomatic tensions and fears of military conflict (Taha, 2019).

In the Tigris-Euphrates basin, Turkey, Syria, and Iraq share water resources, but Turkey's large-scale dam projects, such as the Atatürk Dam, have reduced the flow of water downstream to Syria and Iraq. This has sparked disputes over water rights and led to calls for equitable water-sharing agreements. Water scarcity in this region is compounded by other security challenges, including political instability, ethnic tensions, and civil conflicts, making water disputes particularly volatile (Elhadi et al., 2020).

3.1.5.6.3 The risk of armed conflict

As water scarcity exacerbates regional competition, it often becomes entangled with other issues, such as territorial disputes, economic inequalities, and political power struggles. In some cases, control over water resources is seen as an extension of territorial sovereignty, further intensifying conflicts. When diplomatic negotiations fail, and water shortages become acute, the potential for armed conflict increases. A well-known example is the long-standing water conflict between Israel and its neighbors over the Jordan River and other shared aquifers. Control of water resources has been a significant issue in the Israeli-Palestinian conflict, where water access and distribution are linked to broader territorial and political disputes (Selby, 2013).

In addition to inter-state conflicts, water scarcity can also exacerbate internal tensions, particularly in countries where water resources are unequally distributed among regions or communities. For example, in countries like Yemen, where water resources are limited, disputes over water allocation have fueled internal conflicts and insurgencies. The collapse of governance structures in such situations often leads to the militarization of water access, further escalating security risks (Zeitoun & Warner, 2006).

3.1.5.6.4 Climate change and emerging risks

Climate change exacerbates the security risks associated with water scarcity by increasing the frequency and intensity of droughts, floods, and other extreme weather events. Changes in rainfall patterns, coupled with rising temperatures, are expected to further deplete freshwater resources in already water-stressed regions, making water even more contested. In areas such as sub-Saharan Africa and the Middle East, the impacts of climate change on water availability are already contributing to instability, with water becoming a source of both intra- and inter-state conflict (Tacoli, 2019).

As water scarcity worsens due to climate change, competition for water could increase, especially in areas where agricultural and industrial needs drive high water demand. The likelihood of armed conflict over water resources is heightened in such scenarios, where states or non-state actors perceive access to water as critical for survival and security. The situation in Syria, for example, has been partially attributed to a long-standing drought that worsened water shortages and contributed to the social unrest that eventually spiraled into civil war (Kelley et al., 2015).

3.1.5.6.5 The need for regional cooperation and conflict prevention

To mitigate the security risks associated with water scarcity, regional cooperation and diplomatic dialogue are essential. International agreements that promote fair and equitable water-sharing mechanisms are critical to reducing tensions and preventing conflicts. However, such agreements are often difficult to reach, as states prioritize their own national interests, particularly when the stakes are as high as control over freshwater resources.

Efforts to prevent conflict over water must also involve addressing underlying issues of inequality, governance, and development. Countries need to invest in water management technologies, such as desalination and water recycling, and develop comprehensive water policies that promote sustainable water use and equitable distribution. In regions with shared water resources, multilateral institutions, such as the United Nations and the World Bank, play an important role in facilitating negotiations and supporting conflict prevention efforts.

Water scarcity in conflict-prone regions increases the risk of political instability and armed conflict, particularly when control over water resources is tied to national security concerns. Disputes over transboundary water resources, exacerbated by climate change, can escalate tensions between neighboring countries and contribute to broader security risks. Effective diplomacy, cooperation, and investment in water management are critical to addressing the challenges of water scarcity and preventing water-related conflicts.

3.1.5.6.6 Severe agricultural limitations

Water scarcity in MENA also severely impacts the region's **agriculture**. Agriculture in MENA countries accounts for a significant portion of water usage, but the availability of water for irrigation is limited. This has profound effects on food security, employment, and economic stability.

3.1.6 Water scarcity and its impact on agriculture in MENA

Water scarcity in the Middle East and North Africa (MENA) region presents a significant challenge to the agricultural sector, with profound implications for food security, employment, and economic stability. Agriculture in MENA countries relies heavily on water resources, particularly for irrigation, as the region experiences low rainfall and high temperatures, making natural water availability insufficient to meet the needs of crop and livestock production. Consequently, the agricultural sector consumes a disproportionate share of the region's limited water resources, exacerbating the problem of water scarcity.

3.1.6.1 Agricultural dependence on water

In MENA countries, agriculture accounts for a large portion of water usage. For example, in countries like Egypt, Morocco, and Algeria, over 80% of freshwater resources are used for irrigation (FAO, 2020). However, the availability of freshwater in these countries is severely limited. The region is characterized by arid and semi-arid climates, with many countries relying on non-renewable sources, such as groundwater and desalinized water, for irrigation. Over-extraction of groundwater, combined with the lack of adequate replenishment, has led to falling water tables and the salinization of soil, which further reduces the land's productivity (Al-Zu'bi, 2019). The reliance on unsustainable water sources exacerbates the long-term challenges to agricultural sustainability.

3.1.6.2 Implications for food security

The impact of water scarcity on agriculture directly affects food security in the MENA region. As water resources become scarcer, crop yields decline, and the cost of irrigation rises. This leads to reduced agricultural output, particularly for staple crops such as wheat, barley, and rice, which are heavily dependent on irrigation. In countries such as Egypt, which is one of the largest wheat importers in the world, this decline in domestic agricultural production increases dependence on global food markets. As food prices rise due to reduced local supply, food insecurity deepens, making it more difficult for vulnerable populations to access nutritious food (Abu-Erdan & Abdulhadi, 2020). This dynamic also contributes to the rising malnutrition rates in some MENA countries, further exacerbating the region's socio-economic challenges.

3.1.6.3 Economic and employment challenges

Agriculture in MENA countries is a critical source of employment, especially for rural populations. In many countries, agricultural workers make up a significant proportion of the labor force. For instance, agriculture employs around 30% of the workforce in Egypt and around 20% in Morocco (World

Bank, 2018). However, as water scarcity reduces agricultural productivity, it also threatens jobs in the sector. Reduced agricultural output means fewer jobs in farming, processing, and related industries, leading to higher unemployment and underemployment rates, particularly in rural areas. This creates a vicious cycle, where economic instability in the agricultural sector leads to increased poverty and migration, further straining urban areas and exacerbating regional instability.

3.1.6.4 Socio-political consequences

The consequences of water scarcity extend beyond economic and agricultural challenges. In MENA countries, where water scarcity and agricultural dependence are already sources of tension, the situation is further complicated by rapid population growth and the impacts of climate change. The competition for limited water resources often leads to social and political conflicts. In some cases, disputes over water allocation and land use have contributed to civil unrest, as communities and nations fight for access to diminishing resources (Haddad & Matta, 2021). This has implications for regional stability and governance, as states must balance the competing demands of agricultural production, urban growth, and industrial development.

3.1.6.5 Technological and policy solutions

To address these challenges, MENA countries must invest in more efficient and sustainable water management practices. Technologies such as drip irrigation, water-efficient crop varieties, and improved water storage systems can help reduce water wastage in agriculture. Additionally, the implementation of policies that promote water conservation and the equitable distribution of water resources can help mitigate the impacts of water scarcity on the agricultural sector. Furthermore, regional cooperation on shared water resources, such as the Nile and the Tigris-Euphrates rivers, is essential to prevent conflicts and ensure more sustainable management of water supplies.

Water scarcity in the MENA region poses a significant threat to agricultural productivity, food security, and economic stability. The overreliance on irrigation, coupled with dwindling water supplies, has profound consequences for the region's agriculture and the livelihoods of millions of people. To mitigate these impacts, MENA countries must embrace sustainable water management practices, invest in agricultural innovation, and foster regional cooperation on shared water resources to ensure the long-term viability of agriculture and improve food security for the region's growing population.

3.1.6.6 Challenges to agriculture

3.1.6.6.1 Low rainfall and high evaporation

Most countries in the MENA region are arid or semi-arid with high temperatures, leading to high evaporation rates. This reduces the amount of water available for irrigation, especially for crops that require high water inputs. Countries like Egypt, Syria, and Iraq have relied heavily on irrigation systems, but these are increasingly unsustainable due to declining water resources.

3.1.6.6.2 Challenges to agriculture in the MENA region: Low rainfall and high evaporation

Agriculture in the Middle East and North Africa (MENA) faces significant challenges due to the region's predominantly arid and semi-arid climate, characterized by low rainfall and high temperatures. These conditions lead to high evaporation rates, which further reduce the already limited amount of water available for irrigation. In countries like Egypt, Syria, and Iraq, which rely heavily on irrigation for agricultural production, these climatic challenges are increasingly undermining the sustainability of their agricultural systems.

3.1.6.6.3 Low rainfall and its impact on agriculture

The MENA region is one of the driest in the world, with much of its land receiving less than 200 millimeters of rainfall per year (FAO, 2020). This is far below the minimum amount required for rain-fed agriculture, which makes irrigation crucial for food production. However, the lack of regular and sufficient rainfall means that countries in this region must rely on alternative water sources, such as rivers, groundwater, and desalinated water, to meet the water demands of their agricultural sector.

In Egypt, the Nile River is the primary water source for irrigation, but it is facing increasing pressure from upstream countries, such as Ethiopia, with the construction of the Grand Ethiopian Renaissance Dam (GERD), which threatens to reduce the flow of water downstream. Similarly, Iraq and Syria depend on the Tigris and Euphrates rivers for agricultural water, but upstream dams in Turkey have already reduced water flow to these countries. This reduces the availability of freshwater for irrigation, resulting in crop losses, lower agricultural productivity, and a growing dependence on unsustainable water resources.

3.1.6.6.4 High evaporation rates

High temperatures in the MENA region, which can exceed 40°C during summer months, lead to increased evaporation rates. In fact the rate of evaporation often exceeds the amount of rainfall that these countries receive, further exacerbating water scarcity (Al-Zu'bi, 2019). The combination of low rainfall and high evaporation means that the water available for irrigation is often lost to the atmosphere before it can be effectively utilized by crops. This creates an urgent need for more efficient irrigation methods, but many of the irrigation systems in place, such as flood irrigation, are inefficient and waste significant amounts of water.

In countries like Iraq and Syria, the reliance on surface irrigation methods, which involve channeling water across large areas of land, is particularly problematic. These systems waste a significant portion of water, which is not only inefficient but also leads to waterlogging and soil salinization, further reducing the land's productivity over time (FAO, 2020).

3.1.6.6.5 Unsustainability of current irrigation systems

Irrigation has been a cornerstone of agricultural production in the MENA region, especially for crops that require high water inputs, such as wheat, rice, and cotton. However, the declining availability of water for irrigation is making these systems increasingly unsustainable. Groundwater resources, which many countries in the region depend on, are being depleted at an alarming rate. In some areas, groundwater extraction exceeds the natural replenishment rate, leading to a steady decline in water tables and the potential for the long-term collapse of irrigation systems.

In addition, the lack of efficient water management practices and outdated infrastructure in many countries is exacerbating the situation. For instance, in Egypt, where the majority of agricultural production relies on irrigation from the Nile, the infrastructure is aging and in need of modernization. Water loss due to leakage, inefficient water use, and evaporation from open canals contributes to the depletion of already scarce resources (Al-Zu'bi, 2019).

3.1.6.6.6 Economic and food security implications

The implications of low rainfall, high evaporation rates, and unsustainable irrigation systems on agriculture are far-reaching. As water becomes more scarce, agricultural production declines, leading to reduced food supply and higher prices. This poses a threat to food security, particularly in countries like Egypt, where agriculture is a critical component of the economy and livelihood for millions of people. Furthermore, agricultural workers in the region—many of whom are dependent on irrigation-based farming—are at risk of losing their jobs as crops fail due to water scarcity. This leads to increased unemployment, poverty, and migration, further compounding the region's socio-economic challenges.

3.1.6.6.7 Solutions and adaptation strategies

To address these challenges, MENA countries must adopt more efficient water management practices, such as drip irrigation, which minimizes water wastage and targets the root zone of crops. The use of treated wastewater and desalinated water for irrigation could also help alleviate some of the pressure on freshwater sources. Moreover, there is a need for investment in research and development to identify drought-resistant crop varieties that can thrive in arid conditions.

Countries should also prioritize improving irrigation infrastructure by repairing and modernizing canals, implementing water-saving technologies, and promoting policies that encourage the efficient use of water. Additionally, regional cooperation on shared water resources could help mitigate conflicts and ensure the equitable distribution of water for agricultural purposes.

Agriculture in the MENA region faces significant challenges due to the combination of low rainfall, high evaporation rates, and the unsustainable use of water resources for irrigation. As the region grapples with these challenges, it must implement more sustainable water management practices and explore alternative water sources to ensure the long-term viability of its agricultural sector and protect food security for millions of people. Failure to do so could result in economic instability, increased poverty, and social unrest, further exacerbating the region's existing vulnerabilities.

3.1.6.6.8 Decreasing groundwater levels

In many countries, groundwater, which is crucial for irrigation, is being over-extracted at unsustainable rates. This is especially true in countries like Saudi Arabia, which has no surface water resources and relies on deep aquifers for irrigation (Fahmy, 2019).

3.1.6.6.9 Decreasing groundwater levels in the MENA region

Groundwater is an essential water source for agriculture in the Middle East and North Africa (MENA) region, where surface water is limited due to arid climates and low rainfall. In many countries, particularly those without significant surface water resources, groundwater is the primary source of water for irrigation. However, the over-extraction of groundwater is causing alarming declines in water tables, leading to unsustainable water use and long-term consequences for agricultural production and food security.

3.1.6.6.10 Over-extraction and its impacts

Countries like Saudi Arabia, which lack sufficient surface water sources, rely heavily on deep aquifers for their water needs. The country's agriculture sector, which accounts for a large portion of water consumption, depends primarily on these groundwater reserves for irrigation. Unfortunately, this reliance has led to the over-extraction of groundwater at rates far exceeding its natural replenishment. According to Fahmy (2019), Saudi Arabia extracts groundwater at a rate that far surpasses the aquifers' ability to recharge, a process exacerbated by the lack of effective water management and conservation strategies.

As a result, the water levels in aquifers are steadily decreasing, a phenomenon that has been observed not only in Saudi Arabia but in several other MENA countries, including Iraq, Jordan, and Egypt. The extraction of groundwater has led to the phenomenon of "mining" groundwater—where water that was stored for millennia is being consumed faster than it can be replenished. This over-extraction is contributing to the depletion of vital aquifers, reducing the availability of water for future generations (Fahmy, 2019).

3.1.6.6.11 The economic consequences

The unsustainable extraction of groundwater is not only a concern for water availability but also has profound economic implications. As water tables drop, the cost of pumping water increases, especially in regions where deep aquifers are the main source. This raises the operational costs for farmers, who are already facing increasing pressure due to water scarcity. Moreover, as aquifers are depleted, the quality of groundwater may deteriorate, leading to issues such as salinity intrusion, making the water unsuitable for agricultural use.

For countries like Saudi Arabia, which rely heavily on their groundwater resources to support their agricultural sector, the long-term economic stability of the entire country could be at risk if current practices continue. With water becoming scarcer and more expensive, the agricultural output could decline, leading to food security issues and a greater reliance on food imports. This could also exacerbate unemployment in rural areas where agriculture is a primary livelihood (Fahmy, 2019).

3.1.6.6.12 Environmental and social implications

The depletion of groundwater resources also has environmental consequences. Over-extraction of groundwater can lead to land subsidence, where the ground sinks due to the loss of water in the aquifers. In some parts of the MENA region, this has led to the collapse of infrastructure and damage to agricultural land. Additionally, the reduction of water levels in aquifers can cause the contamination of water supplies due to the intrusion of saline water, further compounding the region's water scarcity problems.

Socially, the decreasing availability of groundwater is affecting rural communities that rely on irrigation for their livelihoods. As farmers struggle with declining water resources, there may be increased competition for water, which can lead to conflicts and migration. In countries such as Yemen and Syria, which are already experiencing political instability, the scarcity of water resources can exacerbate existing tensions and contribute to social unrest (Fahmy, 2019).

3.1.6.6.13 Solutions and adaptation strategies

To mitigate the consequences of decreasing groundwater levels, countries in the MENA region must implement more sustainable water management practices. This includes investing in water conservation technologies, such as drip irrigation and water-efficient cropping systems, to reduce the overall demand for water in agriculture. Additionally, countries should focus on improving the management and monitoring of groundwater resources, incorporating policies that limit over-extraction and promote the replenishment of aquifers.

The use of alternative water sources, such as treated wastewater and desalinated water, could also help alleviate pressure on groundwater reserves. Furthermore, regional cooperation on shared water resources, including groundwater, is essential to ensure the equitable and sustainable use of these critical resources.

Decreasing groundwater levels in the MENA region present a major challenge to agricultural sustainability and water security. The over-extraction of groundwater for irrigation is unsustainable and is leading to the depletion of vital aquifers, which has significant economic, environmental, and social consequences. To address this challenge, it is crucial that MENA countries adopt more efficient water management practices, invest in water conservation technologies, and explore alternative water sources. Without these efforts, the region risks facing long-term water shortages and increased social instability.

3.1.6.6.14 Climate change

Climate models predict that water availability in the region will continue to decline due to increased temperatures and reduced rainfall. This will exacerbate existing agricultural limitations, reducing crop yields and threatening food security.

3.1.6.6.15 Climate change and water availability in the MENA region

Climate change poses a significant threat to the already scarce water resources in the Middle East and North Africa (MENA) region. As climate models predict, the region is expected to experience a rise in temperatures and a decrease in rainfall, both of which will further exacerbate the region's water scarcity challenges. These changes will have profound implications for agriculture, which heavily depends on reliable water supplies for irrigation. The combination of rising temperatures and shrinking rainfall will make it even more difficult to maintain sustainable agricultural practices, thereby threatening food security and economic stability in the region.

3.1.6.6.16 Increased temperatures and water loss

Rising temperatures are a key driver of water scarcity in the MENA region. Higher temperatures lead to increased evaporation rates, which reduces the amount of water available for irrigation and exacerbates the natural drying of rivers, lakes, and groundwater reserves. In countries such as Saudi Arabia, Iraq, and Egypt, where temperatures already exceed 40°C during the summer months, this evaporation exacerbates the loss of water from existing freshwater resources, including both surface water and groundwater. According to the Intergovernmental Panel on Climate Change (IPCC), the region is likely to experience an increase in extreme heat events, further stressing water resources (IPCC, 2021).

As the temperature rises, the water cycle is also altered, leading to a decrease in rainfall. In many parts of the MENA region, particularly in the Mediterranean and desert areas, rainfall is already low and irregular. The predicted decline in precipitation will intensify the region's water crisis by reducing the availability of water for both drinking and agricultural purposes. Areas that rely on rainfall for replenishing groundwater supplies may experience faster depletion of these aquifers, compounding the existing pressure on water resources (Elhadi et al., 2020).

3.1.6.6.17 Reduced rainfall and agricultural limitations

The reduction in rainfall directly impacts agricultural productivity in the MENA region. Agriculture is already highly dependent on irrigation due to the region's naturally dry climate, and with water becoming even more scarce, the region faces the risk of declining crop yields. Crops such as wheat, maize, and rice, which are staples in many MENA countries, require significant water for irrigation. As water availability decreases, these crops may become less viable, leading to reduced food production and increased reliance on food imports.

In countries like Syria and Iraq, where agriculture is already struggling due to water scarcity, reduced rainfall will further constrain farmers' ability to grow crops. For instance, in Syria, the ongoing conflict has already led to the destruction of irrigation infrastructure, and the additional pressures from climate change will further strain the agricultural sector (Al-Ansari, 2020). As rainfall decreases, the availability of water for irrigation will be limited, and without alternative sources such as desalinated water or improved water management, food security will be significantly threatened.

3.1.6.6.18 Impacts on food security

The combination of reduced water availability and declining crop yields threatens food security across the MENA region. In addition to the direct impacts on agriculture, climate change is also contributing to rising food prices, as water scarcity limits the ability to produce food domestically. Countries in the MENA region, such as Egypt and Lebanon, which already face challenges in food production due to their reliance on imported food, will be further burdened by higher food prices and reduced availability.

The food security situation will be most dire for vulnerable populations, including the poor and rural communities, who are more dependent on local agricultural production for their livelihoods. These populations will be the first to feel the effects of reduced crop yields, limited water resources, and rising food prices. Furthermore, the increasing uncertainty in agricultural production may lead to social unrest, as seen in some regions of the MENA during times of food crises (Al-Ansari, 2020).

3.1.6.6.19 Adapting to climate change

To mitigate the impacts of climate change on water resources and agriculture, the MENA region needs to implement adaptive strategies that can help cope with reduced water availability. This includes investing in water-efficient agricultural practices, such as drip irrigation and the use of drought-resistant crop varieties. Additionally, countries in the region should explore alternative water sources, including wastewater recycling and desalination, although these solutions come with their own challenges, as discussed earlier.

Climate adaptation also requires a focus on improving water governance and regional cooperation. Shared water resources, such as the Nile, Tigris, and Euphrates rivers, require collaborative management to ensure equitable distribution of water. Developing water-sharing agreements and adopting more sustainable water management practices can help reduce tensions and enhance resilience to climate change impacts.

Climate change presents a formidable challenge to the MENA region's already limited water resources. Rising temperatures, reduced rainfall, and changing weather patterns will exacerbate water scarcity and limit the region's agricultural productivity, threatening food security and economic stability. To address these challenges, countries in the MENA region must invest in sustainable water management practices, adopt climate-resilient agricultural strategies, and improve regional cooperation on water sharing. Without these measures, the region risks experiencing greater socio-economic instability and a worsening water crisis.

3.1.6.7 Impacts on agriculture

The reduction in available water for agriculture in the MENA (Middle East and North Africa) region poses significant threats to food security. Water is a critical resource for agriculture in this region, which relies heavily on irrigation due to its arid and semi-arid climate. As water becomes scarcer, the ability to grow sufficient food diminishes, leading to food shortages, increased prices, and greater reliance on food imports.

3.1.6.7.1 Food security threats

The reduction in available water for agriculture can lead to food shortages, increased food prices, and dependency on food imports. This is particularly concerning for countries like Egypt, where a significant portion of food is produced through irrigation.

3.1.6.7.2 Decreased agricultural productivity

Many MENA countries, such as Egypt, Iraq, and Syria, are already experiencing the consequences of declining water availability for irrigation. In Egypt, for example, the Nile River provides almost all the water for agricultural irrigation, and reductions in water flow due to upstream projects like the Grand Ethiopian Renaissance Dam (GERD) are threatening the stability of Egypt's food production. The country is highly dependent on its irrigation systems to produce crops like wheat, rice, and vegetables. With decreasing water availability, these crops may become harder to sustain, leading to lower yields and a diminished ability to meet domestic food demands (Khalil & Khalil, 2020).

As water scarcity increases, countries will face growing food insecurity, especially in urban areas where the population relies heavily on imported food. The cost of food is likely to rise, driven by both lower domestic production and increased demand for food imports. In many MENA nations, food imports account for a substantial portion of the national diet, and as water shortages reduce local agricultural output, these countries will become even more dependent on global food markets, which are themselves vulnerable to climate change and geopolitical disruptions (Elhadi et al., 2020).

3.1.6.7.3 Economic consequences and rising prices

Food security threats are not only a social issue but also an economic one. As agricultural productivity declines due to limited water resources, countries will face increased food prices, further straining households that already experience economic hardships. The situation is particularly critical for low-income populations, who spend a large portion of their income on food. For instance, in countries like Lebanon and Jordan, where a large proportion of the population is economically vulnerable, rising food prices will exacerbate poverty and increase social inequality (Al-Ansari, 2020).

In addition, with limited domestic agricultural production, MENA countries will be forced to rely more on food imports. However, global supply chains are also under strain due to climate change, political instability, and economic factors. This could lead to a situation where MENA countries face not only rising food prices but also possible supply disruptions, making food even more inaccessible for those who are already struggling (FAO, 2018).

3.1.6.8 The role of irrigation and water management

Irrigation is the cornerstone of agricultural production in the MENA region, but it is becoming increasingly unsustainable due to water scarcity. Countries like Saudi Arabia, which rely on deep groundwater reserves for irrigation, face the depletion of these non-renewable resources. Over-extraction of groundwater has led to declining water tables, and this trend is likely to continue if current practices are not changed. Sustainable water management practices, such as improved irrigation efficiency and the use of wastewater for irrigation, will be crucial to mitigating the impacts on agriculture and food security (Fahmy, 2019).

In addition to improving water use efficiency, the region must also focus on adopting climate-resilient farming techniques, such as drought-resistant crops and rainwater harvesting. These measures can help stabilize food production in the face of water shortages. However, the widespread adoption of such practices requires substantial investment in research and infrastructure, as well as support from government policies and international aid (Khalil & Khalil, 2020).

3.1.6.8.1 Long-term implications

The long-term implications of reduced water for agriculture in the MENA region could be profound. Food shortages and rising prices may lead to greater social unrest, as seen in previous instances where food crises sparked protests and instability. For example, in 2007-2008, food price hikes led to protests in Egypt and other MENA countries, highlighting the deep connection between water, food, and social stability (FAO, 2018).

Furthermore, as the region's agricultural sector declines, it may result in increased migration as rural populations are forced to leave their farms in search of better opportunities in urban areas or abroad. This rural-urban migration will put additional pressure on already overburdened cities and may create new challenges in terms of unemployment, housing, and public services.

The reduction in available water for agriculture in the MENA region is a major threat to food security. As countries face declining agricultural productivity, food shortages, and rising prices, the impacts on economic stability and social peace will become increasingly significant. Governments must prioritize sustainable water management, invest in climate-resilient agricultural practices, and improve water governance to mitigate these risks and ensure long-term food security for the region's growing population.

3.1.6.9 Rural livelihoods at risk

A large portion of the population in rural areas depends on agriculture for their livelihoods. Water scarcity directly impacts these livelihoods, leading to migration and economic hardship for farming communities (Haddad et al., 2017).

3.1.6.9.1 Rural livelihoods at risk: The impact of water scarcity on agricultural communities

Water scarcity in the MENA (Middle East and North Africa) region is a significant threat to the livelihoods of rural populations, who rely heavily on agriculture for their income and sustenance. As water availability declines, farming becomes increasingly unsustainable, leading to economic hardship for rural communities. These impacts often result in forced migration, as families and communities search for alternative sources of income and better living conditions.

3.1.6.9.2 Economic hardship and migration

In many rural areas of MENA, agriculture forms the backbone of the economy, providing employment for a large portion of the population. However, as water resources dwindle, crop yields decrease, and the costs of irrigation increase. In countries like Iraq, Syria, and Egypt, which rely heavily on irrigation systems to grow staple crops like wheat and rice, water scarcity disrupts these essential food production systems (Haddad et al., 2017). Reduced agricultural productivity leads to decreased income for farmers, which often forces them to seek alternative livelihoods.

The economic hardship caused by water scarcity has led to increasing rural-to-urban migration. Families, unable to sustain their livelihoods through farming, move to cities in search of work. This migration exacerbates challenges in urban areas, where overcrowding, inadequate infrastructure, and limited job opportunities strain existing resources. Additionally, the outflow of rural populations disrupts traditional rural economies and accelerates the decline of agricultural industries, further undermining food security and national economies (Haddad et al., 2017).

3.1.6.9.3 Social and economic implications for farming communities

The social and economic implications of water scarcity are particularly dire for small-scale farmers, who often lack the resources to adapt to changing conditions. Unlike large agricultural enterprises, smallholders have limited access to technologies that could improve water efficiency or facilitate alternative farming practices. The absence of sufficient water for irrigation affects not only crop yields but also the viability of livestock farming, which many rural communities depend on for income and food (Al-Ansari, 2020).

As agriculture becomes less profitable, communities face rising poverty levels, and many farmers are unable to afford the necessary inputs such as seeds, fertilizers, and labor. This perpetuates a cycle of poverty and exacerbates food insecurity, making it difficult for rural populations to meet basic needs. The decline of agricultural livelihoods in rural areas also undermines cultural traditions tied to farming practices, further eroding social cohesion and community resilience.

3.1.6.9.4 Gendered impacts

Water scarcity disproportionately affects women in rural areas, particularly in agricultural communities. In many parts of MENA, women are primarily responsible for household water collection, which can become more time-consuming and physically demanding as water resources become scarcer. Women also play a central role in managing small-scale farming operations, including irrigation and food processing. As water availability declines, their workload increases, and the stress of managing household needs alongside agricultural production can lead to a deterioration in their well-being (Elhadi et al., 2020).

Moreover, women in rural areas often have fewer opportunities for economic diversification, leaving them more vulnerable to the impacts of water scarcity. In many cases, migration may lead to the displacement of entire families, with women and children bearing the brunt of the social and economic consequences. This further limits women's ability to access education, healthcare, and other essential services (FAO, 2018).

3.1.6.9.5 Long-term consequences for rural communities

In the long term, the cumulative effects of water scarcity on rural livelihoods are likely to contribute to a persistent rural-urban divide. As more rural families migrate to cities, the demand for urban housing, services, and jobs will intensify. This will lead to increased urban poverty, informal settlements, and social tensions as urban infrastructure struggles to keep pace with population growth (Haddad et al., 2017). Additionally, rural communities that are left behind will continue to face economic decline, with reduced agricultural output leading to a weakened local economy and increased vulnerability to external shocks such as food price volatility.

The impacts on rural livelihoods will also affect the broader economy, as agriculture plays a critical role in many MENA countries' GDP and employment. The shift away from agriculture could result in the loss of rural economic potential, limiting the development of rural industries such as agribusiness, food processing, and tourism. Therefore, addressing water scarcity in rural areas is not only essential for safeguarding livelihoods but also for ensuring sustainable economic development in the region.

Water scarcity in the MENA region represents a significant challenge to rural livelihoods, threatening food security, economic stability, and social cohesion. As agriculture becomes increasingly unsustainable due to declining water resources, rural populations face rising poverty and migration. The impacts of water scarcity on rural livelihoods are multifaceted, with economic, social, and gendered consequences that require urgent attention from policymakers. To address these challenges, it is crucial to promote sustainable water management practices, invest in climate-resilient farming techniques, and support rural communities in diversifying their livelihoods.

Water scarcity in the MENA region presents a complex set of challenges. The reliance on desalination is a costly and energy-intensive solution, raising environmental and economic concerns. Conflict over shared water resources remains a critical issue, with political and security risks escalating as competition for water intensifies. Additionally, agriculture—a key sector in many MENA countries—faces severe limitations due to dwindling water resources, which poses risks to food security and rural livelihoods. Addressing these challenges requires not only technological innovation, such as more efficient water management and alternative irrigation methods, but also regional cooperation to ensure equitable access to shared water resources and to mitigate the impacts of water scarcity on food production and economic stability.

3.2 Sub-Saharan Africa

Sub-Saharan Africa is another region where extreme water scarcity is widespread, exacerbated by rapid population growth, poor infrastructure, and climate change. Although the region is rich in natural resources, including large river basins like the Nile, Zambezi, and Niger, uneven distribution of water and the lack of access to clean water are critical issues. Many countries in the region rely heavily on groundwater, but over-extraction has led to a depletion of this resource in several areas (FAO, 2020).

The region's vulnerability to water scarcity is compounded by its growing population. According to the United Nations, Sub-Saharan Africa's population is projected to double by 2050, leading to increased demand for water. However, inadequate infrastructure and poor water management practices prevent equitable distribution of resources. Additionally, Sub-Saharan Africa faces the consequences of climate change, with many regions experiencing reduced rainfall and more severe droughts, which aggravate water stress (Bates et al., 2008).

3.2.1 Comparative overview of water scarcity in Sub-Saharan Africa

Water scarcity in Sub-Saharan Africa (SSA) is a critical issue impacting the region's economic development, social stability, and environmental health. Due to a combination of climatic factors, rapid population growth, and underdeveloped infrastructure, many countries in the region face severe water stress. This comparative overview explores the varying impacts of water scarcity across Sub-Saharan Africa, examining the severity of water stress in different countries and the approaches used to manage this crisis.

3.2.2 Water scarcity in Sub-Saharan Africa: Overview and regional variability

Sub-Saharan Africa is characterized by a high degree of water stress, with significant variations in water availability and distribution across countries. The region's population growth, coupled with increasing demand for water in agriculture, industry, and urban areas, exacerbates the already strained water resources. Water scarcity in SSA is primarily driven by factors such as climate change, over-extraction, population pressure, and poor water governance (UNEP, 2017). Countries within the region face different levels of water availability, and solutions to water scarcity vary depending on geographical, economic, and social factors.

3.2.3 Key factors contributing to water scarcity in sub-Saharan Africa

3.2.3.1 Climate change

Shifting rainfall patterns, increased evaporation, and extreme weather events like droughts and floods intensify water scarcity, particularly in semi-arid regions (UNEP, 2018).

Climate change is one of the key drivers exacerbating water scarcity in Sub-Saharan Africa. The region is already prone to variable and unpredictable rainfall patterns, and climate change has intensified these variations. Shifting rainfall patterns result in prolonged dry spells in some areas and heavy rainfall in others, which further complicates water availability (UNEP, 2018). In semi-arid regions, increased evaporation due to rising temperatures significantly reduces the amount of water in rivers, lakes, and reservoirs, making it more difficult to meet agricultural, domestic, and industrial water needs. Extreme weather events such as floods and droughts, which have become more frequent due to climate change, also play a significant role in disrupting water supply, causing both short-term shortages and long-term damage to water infrastructure.

3.2.3.2 Population growth

Sub-Saharan Africa's rapidly growing population, which is expected to double by 2050, further exacerbates demand for water (UNDP, 2018).Sub-Saharan Africa is experiencing rapid population growth, which places increased demand on already limited water resources. As the population grows, the demand for water for agricultural production, drinking, sanitation, and industrial use rises exponentially. In urban areas, the expansion of informal settlements puts additional pressure on local water resources and infrastructure, further exacerbating water scarcity. Rural areas, which rely heavily on rain-fed agriculture and groundwater, are particularly vulnerable, as increased demand for water often outstrips the available supply.

3.2.3.3 Pollution and contamination of water sources

Water contamination, caused by poor sanitation practices, industrial pollution, and agricultural runoff, is another critical factor contributing to water scarcity in the region. Contaminated water sources, such as rivers and wells, not only limit access to clean water but also pose significant health risks. In many parts of Sub-Saharan Africa, especially in rural and peri-urban areas, water treatment facilities are either lacking or inefficient, which exacerbates the problem (UNICEF, 2020). The contamination of water supplies further complicates efforts to address water scarcity, as the focus shifts from improving access to water to addressing public health and water purification issues.

3.2.3.4 Deforestation and land degradation

Deforestation and land degradation are contributing to reduced water retention in the environment. Trees and vegetation play a key role in maintaining the water cycle by facilitating rainfall and preventing soil erosion, which in turn helps retain water in the soil. However, widespread deforestation, often driven by agricultural expansion, illegal logging, and urbanization, reduces the natural capacity of the land to absorb and retain water, leading to higher runoff and reduced groundwater recharge (Molden et al., 2010). This not only exacerbates water scarcity but also affects the overall resilience of ecosystems and communities to climate change.

3.2.3.5 Poor water management and infrastructure

Inefficient water management practices and inadequate infrastructure are key challenges in Sub-Saharan Africa, preventing equitable and sustainable access to water. Many regions lack modern water storage systems pipelines, and irrigation infrastructure, which are essential for capturing and distributing water efficiently. Inadequate maintenance of existing water systems, coupled with rapid population growth and urbanization, results in significant water loss and limited access to clean water, especially in rural areas (Bonsu et al., 2019). In addition, poor governance, corruption, and weak regulatory frameworks often result in the mismanagement of water resources, further exacerbating the scarcity.

3.2.3.6 Conflict and political instability

Political instability and armed conflict also contribute to water scarcity in Sub-Saharan Africa. Conflicts disrupt water supply systems, damage infrastructure, and displace populations, making it more difficult for communities to access water. In some cases, competition over water resources, especially in transboundary river basins, can lead to tensions between countries. For example, disputes over shared water resources, such as the Nile River, have been sources of tension between countries in the region, including Ethiopia, Egypt, and Sudan (Hassan et al., 2021). Such conflicts not only hinder water access but also exacerbate the broader socio-economic challenges facing affected countries.

Water scarcity in Sub-Saharan Africa is driven by a complex interplay of factors, including climate change, population growth, pollution, deforestation, poor water management, and political instability. Addressing these challenges requires a multi-faceted approach that includes improving water management, investing in infrastructure, promoting sustainable land use practices, and strengthening governance systems. Climate adaptation strategies that focus on enhancing water availability and improving resilience to extreme weather events will be critical to ensuring a secure water future for the region.

3.2.3.7 Over-extraction of water resources

Groundwater and surface water resources are being exploited faster than they can be replenished, with agriculture being the largest water consumer (Rost et al., 2019).

Over-extraction of water resources is a significant challenge in many parts of the world, particularly in regions such as Sub-Saharan Africa, the Middle East, and South Asia. Groundwater and surface water resources are being exploited at unsustainable rates, often exceeding their natural replenishment capacity. This is particularly true in agriculture, which is the largest consumer of water globally. According to Rost et al. (2019), the agricultural sector uses approximately 70% of all freshwater resources worldwide, primarily for irrigation. In regions with limited surface water availability, groundwater has become a critical source for irrigation and domestic use. However, the extraction of groundwater at rates faster than it can be replenished leads to the depletion of aquifers, which in turn results in a decline in water availability for both current and future generations.

The over-extraction of water also poses several long-term environmental and economic risks. For instance, over-extraction can lead to the drying up of rivers, lakes, and wetlands, which are vital ecosystems for biodiversity and human livelihoods. Additionally, the excessive use of groundwater can lead to land subsidence and the intrusion of saltwater into freshwater aquifers, making the water unsuitable for consumption or irrigation. This situation is exacerbated by climate change, which is expected to alter precipitation patterns, reduce water availability, and further stress already over-exploited water resources.

In agricultural regions, the increasing reliance on groundwater for irrigation has led to a cycle of water depletion that threatens food security and agricultural productivity. Countries like India, Pakistan, and parts of Sub-Saharan Africa have experienced significant declines in groundwater levels, which has led to increased costs for farmers as they need to drill deeper wells to access water. Furthermore, over-extraction in certain regions, such as the North China Plain and parts of the United States, has caused a drop in groundwater levels by several meters per year, which is unsustainable in the long term (Rost et al., 2019).

To address this issue, it is crucial to implement more sustainable water management practices, including improving irrigation efficiency, using alternative water sources like treated wastewater, and enhancing water storage and conservation efforts. Additionally, policies should be enacted to regulate the extraction of groundwater and surface water resources to prevent overexploitation and ensure long-term water availability.

3.3 Country-Specific Impacts of Water Scarcity in Sub-Saharan Africa

While water scarcity is a regional challenge, its intensity and manifestations differ across countries in Sub-Saharan Africa. Below are some examples of countries in the region that experience water scarcity, highlighting the severity of the issue and the specific impacts.

3.3.1 Ethiopia: Vulnerability to drought and groundwater stress

Ethiopia, one of the most water-stressed countries in the Horn of Africa, relies heavily on rainfall for water resources. The country's major water bodies include the Blue Nile and Rift Valley lakes, but many areas suffer from erratic rainfall, seasonal droughts, and groundwater depletion.

Ethiopia is one of the most water-stressed countries in the Horn of Africa, with its water resources significantly influenced by the variability of rainfall patterns. The country's major water bodies, such as the Blue Nile River and the Rift Valley lakes, are crucial for agriculture, domestic consumption, and hydropower generation. However, Ethiopia faces several challenges related to water scarcity due to erratic rainfall, seasonal droughts, and groundwater resource depletion, exacerbated by climate change and unsustainable water management practices.

Rainfall in Ethiopia is highly variable, with some regions experiencing heavy seasonal rains, while others, particularly in the eastern and southeastern parts of the country, face prolonged dry spells. These erratic rainfall patterns significantly affect the availability of surface water resources, which many Ethiopian communities rely on for drinking, irrigation, and livestock. The irregularity of rainfall, combined with high evaporation rates, means that some areas experience acute water shortages, especially during drought periods (Taddese et al., 2018). These droughts, which have become more frequent and severe due to climate change, lead to reduced water availability for agriculture, further threatening food security and economic stability in rural areas.

In addition to surface water stress, Ethiopia is also facing groundwater depletion. Groundwater is an important source of water, particularly in regions where surface water is scarce. However, the over-extraction of groundwater for irrigation and domestic use, without adequate replenishment, has led to declining water levels in many aquifers. This over-reliance on groundwater, especially in areas like the Awash Basin and parts of the Rift Valley, places significant pressure on already limited resources (Alemayehu & Awulachew, 2020). Groundwater stress is further compounded by poor water management practices, lack of infrastructure for sustainable extraction, and insufficient investment in water conservation techniques.

The combined effects of erratic rainfall, drought, and groundwater depletion have made Ethiopia highly vulnerable to water scarcity. This vulnerability is particularly concerning as agriculture, which accounts for a large portion of the country's economy and employment, heavily depends on consistent water availability. As water resources become increasingly scarce, Ethiopia faces the dual challenge of managing its water systems more effectively while adapting to the impacts of climate change.

To mitigate these challenges, Ethiopia needs to invest in more sustainable water management strategies, such as improving water storage, increasing the efficiency of irrigation systems, and expanding the use of rainwater harvesting. Furthermore, efforts to protect and manage groundwater resources, coupled with better monitoring and regulation of water extraction, are essential to ensure long-term water security.

3.3.1.1 Challenges

The country is highly vulnerable to droughts and climate variability, which affect the already limited water resources. Rural areas are particularly hard-hit, with agriculture and livestock farming being the primary means of livelihood (Bachewe et al., 2020).

Ethiopia faces significant challenges related to water scarcity, exacerbated by its vulnerability to droughts and climate variability. As one of the most water-stressed countries in the Horn of Africa, Ethiopia's water resources are already limited, and these resources are under increasing pressure due to fluctuating rainfall patterns and prolonged dry spells. The country's reliance on rainfall, which is highly variable, has made it particularly susceptible to the impacts of climate change, including more frequent and intense droughts (Bachewe et al., 2020). These changing climatic conditions are destabilizing agricultural and water systems, which are central to the livelihoods of rural communities.

In rural Ethiopia, agriculture and livestock farming are the primary means of livelihood for a large portion of the population. These sectors are particularly vulnerable to water scarcity, as crop production and grazing for livestock are heavily dependent on reliable water sources. When water availability decreases due to drought or shifting rainfall patterns, crop yields suffer, and livestock face food and water shortages. This results in reduced agricultural productivity, increased poverty, and food insecurity. Many rural areas, where communities are already living with limited access to clean and potable water, are hit hardest by these challenges. The lack of reliable water sources also affects the health and sanitation conditions in these areas, further exacerbating the challenges of water scarcity.

The impact of droughts on Ethiopia's rural population extends beyond agriculture, as water scarcity also affects access to drinking water and sanitation. In areas where wells and rivers dry up or water quality deteriorates, communities are forced to travel longer distances to collect water, often from unsafe sources, which poses a significant health risk. In this context, the dual pressures of environmental stress and socio-economic vulnerability heighten the challenges faced by rural populations, leaving them more exposed to the effects of climate change and water scarcity.

Ethiopia needs to invest in climate-resilient water infrastructure to address these challenges, such as improved irrigation systems, rainwater harvesting techniques, and better groundwater management practices. Additionally, strategies to diversify livelihoods in rural areas, improve food security, and ensure access to safe drinking water are essential for reducing the impact of droughts and climate variability.

3.3.1.2 Solutions

Ethiopia has initiated large-scale irrigation projects and is investing in rainwater harvesting to improve water supply. However, issues with water governance and equitable distribution persist (Fenta et al., 2021).

Ethiopia has recognized the urgent need to address its water scarcity challenges, particularly in rural areas where agriculture and livestock farming are central to livelihoods. In response, the government has launched several large-scale irrigation projects aimed at improving water availability and supporting agricultural production. These projects, such as the construction of irrigation systems in the Awash and Omo River basins, are designed to harness available water more efficiently, increase crop yields, and reduce reliance on unpredictable rainfall patterns (Fenta et al., 2021). Additionally, Ethiopia has increasingly invested in rainwater harvesting systems, which can provide a reliable water supply for both agriculture and domestic use, particularly in arid and semi-arid regions where conventional water sources are scarce.

Despite these efforts, several challenges remain. One of the key issues is water governance. The equitable distribution of water resources has proven to be difficult, particularly given the increasing competition for water among different sectors, such as agriculture, urban areas, and industrial use. There is also the challenge of ensuring that the benefits of these projects are shared fairly across the population, especially among marginalized communities that may lack the infrastructure to benefit from irrigation schemes or rainwater harvesting technologies. This problem is compounded by inconsistent water management policies and a lack of coordination between local and national levels of government, which hinders the effective allocation and use of water resources (Fenta et al., 2021).

Moreover, although large-scale irrigation projects have the potential to boost agricultural productivity, they require substantial financial investment and the expertise to manage complex infrastructure systems. This has raised concerns about the long-term sustainability of these projects, especially in a country where funding for water-related infrastructure is often limited, and climate variability can undermine the effectiveness of irrigation schemes (Bachewe et al., 2020). Without a robust and integrated approach to water management, including a focus on local community involvement and the strengthening of water governance frameworks, the challenges of equitable water distribution and sustainable water usage will persist.

To address these issues, Ethiopia could benefit from adopting a more inclusive approach to water management, ensuring that all stakeholders, including local communities, have a say in water-related decisions. This would involve improving infrastructure and enhancing water governance mechanisms, promoting sustainable practices, and ensuring that water policies align with the needs of the most vulnerable populations.

3.3.2 South Africa: Over-reliance on groundwater and desalination

South Africa, one of the most industrialized countries in SSA, is facing severe water scarcity, especially in urban centers like Cape Town, where droughts have become more frequent. South Africa relies heavily on groundwater and large dams for water supply.

South Africa, one of the most industrialized countries in Sub-Saharan Africa, is facing significant water scarcity, particularly in urban centers like Cape Town, which has experienced several severe droughts in recent years. The country's water supply is heavily dependent on groundwater and large dams, but these resources are increasingly under stress due to over-extraction and the impacts of climate change. Groundwater is particularly important in South Africa's semi-arid regions, where surface water sources are limited, and it serves as a critical backup during periods of drought. However, over-reliance on groundwater has led to the depletion of aquifers, which are often unsustainably extracted at rates faster than they can be replenished.

The situation in Cape Town highlights the vulnerability of South Africa's water systems. In 2018, the city faced "Day Zero," when the city's reservoirs were predicted to reach critically low levels, and water taps might be turned off, affecting millions of residents (Palmer, 2019). This crisis was exacerbated by a combination of factors, including extended droughts, poor water management, and population growth, which increased demand for water. In response to the crisis, the city and the broader region have turned to desalination as an emergency measure to increase water supply. Desalination plants have been established in coastal areas to convert seawater into potable water, providing an additional, though energy-intensive, source of water. However, desalination presents significant challenges, including its high cost, environmental impacts, and dependence on electricity, which in South Africa is largely generated from fossil fuels.

While desalination is seen as a potential solution to water scarcity, it cannot provide a long-term fix for South Africa's water crisis. Groundwater depletion, compounded by climate change-induced reductions in rainfall and increasing evaporation rates, continues to threaten the sustainability of water resources across the country. In the long term, South Africa needs to adopt integrated water management strategies that include better regulation of groundwater use, investment in water recycling technologies, and improvements in water conservation and efficiency practices. Additionally, addressing climate change through mitigation and adaptation strategies will be crucial in reducing the frequency and severity of droughts that exacerbate water scarcity.

South Africa's water scarcity highlights the critical need for more sustainable water management practices. Efforts must focus on balancing water supply with demand, improving infrastructure for water storage and distribution, and ensuring that water resources are used efficiently. Additionally, promoting water conservation at the community level, improving rainwater harvesting techniques, and diversifying water sources through technology and innovation will be key components of a resilient water strategy for the future.

3.3.3 Challenges

South Africa's water demand outstrips supply, with a severe reliance on groundwater exacerbating over-extraction. The country also faces increasing demand from agriculture, industry, and growing urban populations (Pereira et al., 2019).

South Africa faces a critical water scarcity challenge, primarily driven by the growing demand for water in agriculture, industry, and urban areas, which has begun to outstrip the available supply. This issue is compounded by the country's heavy reliance on groundwater resources, which are being over-extracted at unsustainable rates. Groundwater, particularly in areas like the Western Cape, serves as an essential backup water source during drought periods. However, over-extraction has led to the depletion of aquifers, a problem that is made worse by inadequate recharge rates and the impacts of climate change, which alter rainfall patterns and increase evaporation rates (Pereira et al., 2019).

Agriculture is a major consumer of water in South Africa, accounting for around 60% of the country's total water usage. The sector's high water demand, combined with climate change-induced shifts in rainfall patterns, has exacerbated the country's water challenges. Extended dry spells, coupled with the increasing competition for water from urban centers and industries, are putting immense pressure on the already strained water resources (Pereira et al., 2019). The result is an over-exploitation of groundwater resources, particularly in rural areas, where agriculture is the primary livelihood for many communities. This depletion of groundwater is a significant concern as it threatens long-term water security and exacerbates the vulnerability of agricultural production, which in turn affects food security.

The urban population in South Africa is growing rapidly, especially in cities like Cape Town, Johannesburg, and Pretoria, which further intensifies water demand. Urbanization has increased the need for both potable water and water for industrial processes, and existing water supply systems are struggling to keep up with the escalating demand. The combination of a growing population, water scarcity, and strained infrastructure has led to frequent water shortages and rationing, particularly in cities like Cape Town, which faced the threat of "Day Zero" in 2018, when the city's reservoirs were predicted to run dry (Palmer, 2019).

Given these challenges, South Africa faces an urgent need for comprehensive water management strategies. These strategies must focus on reducing water wastage, improving water storage infrastructure, and encouraging sustainable agricultural practices that use water more efficiently. Additionally, the country will need to increase investments in alternative water sources, such as desalination and wastewater treatment and reuse, to diversify its water supply. At the same time, robust governance and regulatory frameworks must be implemented to manage groundwater resources effectively and ensure equitable access to water for all sectors of society.

3.3.4 Solutions

South Africa has implemented desalination plants and water recycling projects, particularly in Cape Town, to manage water resources. However, water conservation and efficient distribution remain a significant challenge (Khumalo & Mazvimavi, 2019).

In response to the growing water scarcity crisis, South Africa has taken steps to manage its limited water resources more effectively. One of the most notable solutions has been the development of desalination plants, particularly in coastal cities like Cape Town, where water shortages have become more frequent due to droughts. Desalination provides an alternative source of freshwater by converting seawater into potable water. During Cape Town's 2018 water crisis, desalination was seen as a critical component of the city's emergency water supply strategy (Khumalo & Mazvimavi, 2019). Similarly, water recycling projects have been introduced in urban areas to treat and reuse wastewater for non-potable uses, such as irrigation and industrial processes.

However, these measures alone are not enough to address South Africa's water challenges in the long term. Despite technological advancements in desalination and water recycling, water conservation and efficient distribution remain significant challenges. South Africa's water systems suffer from inefficiencies in water management, with high levels of water loss due to leaking pipes, outdated infrastructure, and inadequate maintenance. In rural areas, where groundwater resources are critical for irrigation, the over-extraction of aquifers continues to deplete water reserves faster than they can be replenished (Pereira et al., 2019). These inefficiencies exacerbate the water scarcity problem, making it even more difficult to ensure equitable access to water for all communities.

To address these challenges, South Africa must adopt a more integrated approach to water management. This includes enhancing water conservation efforts through public awareness campaigns, incentivizing efficient water usage in agriculture and industry, and improving infrastructure to reduce water losses. Additionally, expanding and improving water distribution systems, especially in underserved rural areas, will help ensure that water is distributed more equitably across the population. Furthermore, improving governance and regulatory frameworks to manage water resources more effectively is essential. This can be achieved through better monitoring of water usage, ensuring that water allocations are made fairly across sectors, and enforcing policies that promote sustainable water practices.

In the long term, South Africa's water strategy must also focus on climate change adaptation, as the country is highly vulnerable to shifts in rainfall patterns and increased temperatures. The implementation of climate-resilient agricultural practices, such as the use of drought-resistant crops and efficient irrigation systems, can help reduce the sector's water consumption and improve food security. These efforts will require coordinated action across all levels of government, the private sector, and local communities to ensure that water resources are managed sustainably and equitably.

3.4 Nigeria: Conflict Over Water Resources and Groundwater Depletion

Nigeria, the most populous country in Sub-Saharan Africa, experiences significant water stress due to a combination of pollution, population pressure, and political conflict over water resources.

Nigeria, as the most populous country in Sub-Saharan Africa, faces critical challenges related to water scarcity, exacerbated by both natural and human-induced factors. The country's vast population of over 200 million people puts immense pressure on available water resources, contributing to issues such as groundwater depletion, inadequate infrastructure, and pollution. Nigeria's water crisis is further complicated by regional disparities in water availability, with the northern parts of the country facing more severe water stress than the southern regions. Additionally, political conflict over the allocation and control of water resources has fueled tensions, especially in areas where water bodies cross-regional or ethnic boundaries.

3.4.1 Groundwater depletion

Groundwater is a key source of freshwater in Nigeria, especially in arid and semi-arid regions. However, over-extraction for agriculture, domestic use, and industrial activities has led to significant depletion of groundwater resources. In some parts of Nigeria, such as the northern states, groundwater tables have dropped dramatically due to unsustainable extraction rates. The country's growing urbanization and increasing demand for water further exacerbate this issue. The lack of comprehensive water management strategies and the absence of sufficient regulation on groundwater extraction contribute to the depletion, which threatens the long-term availability of this vital resource (Olajuyigbe & Taiwo, 2019).

3.4.2 Pollution of water sources

In addition to over-extraction, water pollution is a significant problem in Nigeria. Industrial waste, oil spills, agricultural runoff, and improper waste disposal all contribute to the contamination of both surface water and groundwater. The Niger Delta, for instance, has suffered from extensive oil spills, which have not only caused environmental degradation but also contaminated freshwater sources, making them unsafe for consumption. This pollution is a major threat to the health of local populations, especially in rural communities, where access to clean water is already limited (Sule et al., 2018).

3.4.3 Conflict over water resources

Water-related conflicts are prevalent in Nigeria, particularly in areas where water sources are shared across regions, states, or ethnic groups. In the northern and central regions, where drought and water scarcity are common, competition for access to water resources has fueled tension between communities, especially between farmers and herders. This competition is often aggravated by political dynamics and competition for land and resources. In some cases, disputes over water rights have escalated into violent clashes, further undermining efforts to ensure equitable access to water. These conflicts are often intensified by the broader context of ethnic and political rivalries, which complicate efforts to negotiate water-sharing agreements (Alemayehu & Shishira, 2020).

3.4.4 Solutions and policy recommendations

To address these challenges, Nigeria must adopt an integrated approach to water resource management that balances the needs of various sectors while ensuring equitable access. This includes implementing stronger regulations on groundwater extraction, improving water conservation practices, and investing in sustainable water infrastructure. Additionally, policies to tackle water pollution, such as stricter enforcement of environmental laws and investments in wastewater treatment facilities, are necessary to protect water sources. Nigeria also needs to foster dialogue and cooperation between different stakeholders, including local communities, governments, and industries, to resolve conflicts over water resources. Effective governance, transparent allocation, and equitable distribution of water are key to reducing tensions and ensuring long-term water security for the country (Olajuyigbe & Taiwo, 2019).

3.4.4.1 Challenges

The Niger River, which flows through Nigeria, is shared by several countries, leading to disputes over access. Additionally, many rural areas depend on groundwater, which is increasingly contaminated and depleted due to overuse (Adeyemo, 2020). The northern regions of Nigeria are particularly vulnerable to desertification and drought, which intensify water scarcity.

The Niger River, which is a vital water source for several countries in West Africa, flows through Nigeria and is shared by multiple nations, including Niger, Mali, and Benin. This shared nature of the river has led to significant disputes over water access, with each country vying for a fair share of its resources. For Nigeria, the Niger River is crucial for agriculture, transportation, and domestic water supply, particularly in the middle and lower reaches. However, the competing interests and lack of coordinated management between these countries have created tensions over the river's utilization. While there are agreements like the Niger Basin Authority aimed at managing the river's resources, the enforcement of these agreements remains weak, and disputes over water allocation continue to be a source of conflict (Adeyemo, 2020).

In addition to these geopolitical challenges, rural areas in Nigeria, particularly in the north, face significant difficulties related to groundwater. Groundwater is a crucial water source in many parts of the country, especially in arid and semi-arid regions where surface water is scarce. However, groundwater resources are being severely depleted due to over-extraction for agricultural and domestic use, especially in regions that rely on irrigation. Moreover, the contamination of groundwater from industrial activities, oil spills, and improper waste disposal has compounded the issue. This contamination renders groundwater unsafe for drinking and irrigation, worsening water scarcity and public health concerns (Olajuyigbe & Taiwo, 2019).

The northern regions of Nigeria, which are more susceptible to desertification and drought, face additional challenges in water availability. Desertification, driven by unsustainable farming practices, deforestation, and climate change, reduces the land's ability to retain water, leading to lower water availability for agricultural activities. The intensification of droughts due to changing climate patterns further exacerbates the water crisis in these areas. As a result, communities in northern Nigeria often experience severe water shortages during dry seasons, forcing them to rely on increasingly limited and polluted water sources (Adeyemo, 2020). These environmental factors, coupled with the overuse of groundwater and inadequate water management practices, pose significant barriers to addressing water scarcity in the country.

3.4.4.2 Solutions

Nigeria has launched initiatives to improve rainwater harvesting and irrigation techniques to bolster agricultural productivity and increase water availability. However, political instability complicates water management efforts (Aluko et al., 2020).

To address the growing water scarcity issues, Nigeria has implemented several initiatives aimed at improving water availability, particularly for agricultural purposes. One such initiative involves the promotion of rainwater harvesting, which is seen as a viable solution for increasing the country's water supply, especially in areas with unreliable rainfall patterns. Rainwater harvesting systems have been installed in rural and urban areas to capture and store rainwater, reducing the dependence on surface water and groundwater. Additionally, Nigeria has introduced more efficient irrigation techniques, such as drip irrigation, which reduce water wastage and allow for more sustainable agricultural practices in water-scarce regions (Aluko et al., 2020).

Despite these efforts, Nigeria's water management strategies are often hindered by political instability and weak governance. In regions plagued by insurgency and internal conflicts, such as the northeastern parts of the country, efforts to implement and maintain water management projects are frequently disrupted. For instance, the Boko Haram insurgency has led to the destruction of infrastructure, including water supply systems, making it difficult for affected populations to access safe water (Aluko et al., 2020). Additionally, political infighting and corruption at various levels of government often delay or derail large-scale water management projects, leaving critical water-related infrastructure in disrepair and poorly managed.

Moreover, although Nigeria has introduced policies aimed at improving water governance, including the creation of the National Water Resources Bill, the implementation of these policies has been inconsistent. The lack of coordination between federal, state, and local governments, as well as between different water-related sectors, has created gaps in the management of water resources. Effective water management requires a cohesive, nationwide approach that includes equitable distribution, sustainable extraction, and protection from contamination. Without addressing these institutional weaknesses, Nigeria's efforts to improve water access and prevent further depletion of its resources are likely to remain inadequate (Aluko et al., 2020).

3.5 Kenya: Scarcity of Freshwater and Over-reliance on Water Transferal

Kenya, located in East Africa, experiences water scarcity primarily due to uneven rainfall distribution and increasing demand from agriculture and urban areas. The country's freshwater resources are concentrated in the highland regions, with large parts of the country experiencing arid and semi-arid conditions.

Kenya faces significant water scarcity challenges due to a combination of uneven rainfall distribution, climate variability, and increasing water demand from both agriculture and urban areas. The country's freshwater resources are primarily concentrated in the highland regions, which benefit from more consistent and abundant rainfall. In contrast, much of Kenya, particularly in the northeastern and southeastern regions, experiences arid and semi-arid conditions, with rainfall being sparse and unreliable. These regions, which make up about 80% of the country's land area, struggle with limited access to freshwater resources, contributing to widespread water stress (Omondi et al., 2021).

The situation is compounded by the fact that water demand in Kenya is rising rapidly, driven by population growth, urbanization, and expanding agricultural activities. Agriculture is the largest consumer of water in Kenya, accounting for approximately 70% of the country's total water use. However, this sector is heavily dependent on rainfall, and erratic precipitation patterns exacerbate the vulnerability of smallholder farmers. Irrigation, which could offer a solution to mitigate rainfall dependency, remains underdeveloped due to inadequate infrastructure and high costs, further intensifying the pressure on available water resources (Mogaka et al., 2020).

Urbanization is also placing a significant strain on water resources, particularly in major cities such as Nairobi and Mombasa, where rapid population growth is outpacing the development of water supply infrastructure. In Nairobi, for instance, the demand for water often exceeds supply, leading to frequent water shortages, especially during dry spells. The growing urban population further exacerbates competition for the limited available freshwater, and both urban and rural areas face the challenge of managing water resources equitably (Nyangena et al., 2021).

Furthermore, Kenya is not immune to the impacts of climate change, which is expected to exacerbate water scarcity by altering rainfall patterns, increasing evaporation rates, and intensifying the frequency and severity of droughts. These climate-related shifts have already been observed in recent years, with longer dry periods and shorter rainy seasons leading to reduced water availability. The country has experienced recurrent droughts, particularly in arid and semi-arid regions, which have resulted in crop failures, livestock deaths, and displacement of rural populations. This highlights the urgent need for effective water management policies that address both current and future challenges related to water scarcity (UNEP, 2018).

3.5.1 Challenges

Water availability is highly unequal, with most of Kenya's water resources concentrated in the mountainous and coastal areas, while northern Kenya remains water-poor (Ong'uti et al., 2019). Additionally, population pressure and climate change are expected to further reduce water availability in the coming decades.

Water availability in Kenya is highly unequal, with the majority of the country's freshwater resources concentrated in the highland and coastal regions, while northern Kenya remains water-poor. The country's highland areas, including parts of the central and western regions, receive relatively abundant rainfall, supporting more consistent water supplies. However, large portions of the country, particularly in the north and northeastern regions, are arid or semi-arid, where rainfall is sparse and irregular. This uneven distribution of water resources poses significant challenges for equitable access to water, particularly in the more vulnerable regions (Ong'uti et al., 2019).

In addition to the geographical disparity in water availability, population pressure is rapidly increasing, especially in urban areas like Nairobi and Mombasa. As Kenya's population continues to grow, so does the demand for water, which intensifies competition for the limited water resources available. Urban areas are particularly affected by water scarcity due to inadequate infrastructure, rapid urbanization, and the challenges of managing water supplies in growing cities. As the rural population migrates to urban centers in search of better opportunities, the strain on already overstretched water systems increases, leading to frequent water shortages (UNEP, 2018).

Furthermore, climate change is exacerbating water scarcity in Kenya. Rising temperatures, changing rainfall patterns, and increased frequency of extreme weather events, such as prolonged droughts and flooding, threaten the stability of the country's water resources. In regions that are already water-stressed, such as northern Kenya, climate change is expected to reduce the availability of water for agriculture, livestock, and domestic use. These shifts in weather patterns are predicted to increase the vulnerability of rural communities that depend on rain-fed agriculture and surface water for their livelihoods. The added pressure from climate change underscores the need for adaptive strategies to manage water resources sustainably and reduce vulnerability (Ong'uti et al., 2019).

3.5.2 Solutions

Kenya has invested in water transfer schemes and is looking to develop desalination plants in coastal regions. However, the sustainability of these efforts remains a question due to high operational costs and limited infrastructure (Karanja et al., 2018).

Kenya has recognized the growing challenge of water scarcity and has initiated various solutions to address the issue. One of the key strategies employed is the development of water transfer schemes, which aim to redistribute water from water-rich areas to more arid regions. For example, the government has been working on projects such as the Northern Water Collector Tunnel, designed to divert water from the highland regions to the drought-prone areas in the central and northern parts of the country. This approach seeks to balance the distribution of water resources and ensure a more equitable water supply across regions (Karanja et al., 2018).

In addition to water transfer, Kenya is exploring the potential of desalination as a solution, particularly in its coastal areas. The country has been considering desalination plants to provide a reliable source of freshwater, especially in urban centers like Mombasa, where water demand often exceeds supply. Desalination could offer a sustainable alternative to traditional sources of water; however, the high operational costs and energy demands of desalination technologies pose significant challenges. The need for substantial infrastructure investment and maintenance, along with the environmental impacts of brine disposal, further complicates the feasibility of large-scale desalination in the region (Karanja et al., 2018).

Despite these efforts, the sustainability of Kenya's water solutions remains uncertain. The high costs associated with water transfer projects and desalination plants, as well as the challenges in developing and maintaining the necessary infrastructure, hinder the long-term viability of these initiatives. Moreover, the country faces challenges in improving water storage, increasing conservation efforts, and ensuring equitable water distribution, particularly in rural and marginalized areas. Without a comprehensive approach that addresses both supply and demand, Kenya may continue to struggle with water scarcity in the coming decades (Karanja et al., 2018).

3.6 Cross-Country Comparison of Water Scarcity in Sub-Saharan Africa

Water scarcity varies significantly across Sub-Saharan Africa. Countries like Ethiopia and South Sudan, heavily reliant on rain-fed agriculture, are vulnerable to droughts. Coastal nations like Nigeria and Ghana face challenges from saltwater intrusion and pollution. Urban centers like Lagos and Nairobi experience rapid population growth, leading to increased water demand and infrastructure strain. Climate change exacerbates these issues, with more frequent and intense droughts and floods impacting water availability and quality. Addressing water scarcity requires integrated approaches, including improved water management, investments in infrastructure, and climate adaptation strategies.

3.6.1 Severity of water scarcity

Water scarcity is a severe global crisis, affecting billions of people. It manifests in various forms, including physical scarcity, where water resources are insufficient to meet demand, and economic scarcity, where access to water is limited due to poverty or inadequate infrastructure. The severity of water scarcity varies across regions, with arid and semi-arid areas facing the most acute challenges. Climate change further exacerbates the situation, leading to more frequent and intense droughts and floods. Water scarcity has far-reaching consequences, impacting agriculture, industry, and human health. It can lead to food shortages, economic decline, and social unrest. Addressing water scarcity requires urgent action, including sustainable water management practices, investments in water infrastructure, and climate adaptation measures.

Ethiopia and *South Africa* face severe water scarcity, with urban areas experiencing water crises and rural areas lacking reliable access to clean water.

Ethiopia and South Africa are among the most water-stressed countries in Africa, grappling with severe water scarcity that affects both urban and rural areas. In Ethiopia, water resources are highly dependent on rainfall, and the country faces significant variability in precipitation patterns. This unpredictability, coupled with prolonged droughts, exacerbates water availability issues, particularly in rural regions where communities rely heavily on seasonal rainfall for agricultural activities and domestic use. Urban centers like Addis Ababa experience water crises as demand outstrips supply, and infrastructure challenges make it difficult to ensure consistent water access. Groundwater resources, while an important backup, are being depleted at unsustainable rates, further stressing the system (Bachewe et al., 2020). In South Africa, cities like Cape Town have faced significant water crises, such as the 2018 “Day Zero” crisis, where the city was on the brink of running out of water. The situation in South Africa is compounded by the over-reliance on groundwater and surface water resources that are increasingly strained due to high agricultural demand, climate change, and industrial growth. Water scarcity is becoming a critical issue, especially in urban centers where water management infrastructure struggles to keep up with rapid urbanization (Pereira et al., 2019).

Nigeria and **Kenya** experience moderate water stress, with specific regions within each country suffering from either high population density, pollution, or environmental degradation. Both countries have invested in large-scale water management solutions but still face significant challenges.

Nigeria and Kenya face moderate water stress, where specific regions experience significant water challenges due to factors such as high population density, pollution, and environmental degradation. Nigeria, with its large population, is particularly vulnerable to water scarcity in urban areas like Lagos and Kano, where rapid population growth leads to high water demand. The Niger River, which flows through the country, is shared by several nations, and conflicts over water resources have further complicated water management. Pollution from industrial activities, poor sanitation, and oil spills have severely impacted both groundwater and surface water quality, making clean water scarce for many communities, particularly in the northern and southern parts of the country (Adeyemo, 2020). In rural areas, the over-extraction of groundwater, combined with poor water management, has led to groundwater depletion, further worsening the water crisis. Similarly, Kenya's water stress is region-specific, with the arid and semi-arid northern regions facing chronic water shortages. While Nairobi and Mombasa experience water shortages during periods of drought, they have managed to implement water transfer systems and desalination projects in coastal areas (Karanja et al., 2018). However, these solutions are often costly and require substantial infrastructure investment.

Despite these challenges, both Nigeria and Kenya have taken significant steps to address water scarcity through large-scale water management solutions. Nigeria has focused on improving rainwater harvesting, irrigation techniques, and enhancing water distribution networks, especially in rural and semi-urban areas. However, political instability and insufficient governance frameworks continue to hinder effective implementation (Aluko et al., 2020). In Kenya, large-scale infrastructure projects, including the development of desalination plants in coastal areas and water transfer systems from highland regions to arid areas, have been undertaken to ensure a more equitable distribution of water. However, these solutions come with challenges such as high operational costs, energy demands, and the need for effective maintenance, especially in a country where rural areas remain underserved by reliable water infrastructure (Karanja et al., 2018).

While Ethiopia and South Africa face severe water scarcity, with crises affecting both urban and rural areas, Nigeria and Kenya experience moderate water stress with specific regions facing localized challenges. Both sets of countries have invested in large-scale water management solutions, but ongoing issues such as inadequate infrastructure, climate change, pollution, and over-extraction of water resources continue to undermine efforts to ensure reliable access to clean water. The need for sustainable, equitable, and efficient water management systems remains critical in addressing water scarcity in these countries.

3.7 Management Approaches

Effective water management strategies are crucial to addressing the growing challenge of global water scarcity. As freshwater resources become increasingly strained by population growth, industrial demands, and the impacts of climate change, it is imperative to adopt comprehensive and sustainable approaches to water management. Key strategies such as demand-side management, supply-side management, integrated water resource management (IWRM), and climate-resilient water management are central to ensuring that water resources are used efficiently, equitably, and sustainably.

3.7.1 Demand-side management (DSM)

DSM focuses on reducing water consumption by promoting water-efficient technologies and encouraging conservation practices. This approach includes the adoption of water-saving devices, such as low-flow toilets, efficient irrigation systems, and water-efficient appliances, which can significantly reduce domestic and industrial water usage. Additionally, DSM involves changing consumer behavior through awareness campaigns and education, encouraging individuals and businesses to use water more responsibly. For example, implementing tiered pricing for water, where higher consumption leads to higher costs, can incentivize users to adopt more sustainable water use practices. Policies aimed at promoting water recycling and reuse, particularly in industrial and agricultural sectors, are also essential components of DSM. By addressing the demand side, these strategies can help reduce the overall pressure on water resources, making them available for other critical uses (Gleick, 2014).

3.7.2 Supply-side management (SSM)

SSM focuses on increasing the availability of water through investments in infrastructure, such as reservoirs, pipelines, and treatment plants. This includes building new water storage systems to capture and store water during periods of abundance, which can then be used during dry spells. Additionally, the treatment and distribution infrastructure need to be upgraded to ensure that water is delivered efficiently and with minimal losses. One of the most promising innovations in supply-side management is the use of alternative water sources, such as desalination, which converts seawater into freshwater, and rainwater harvesting, which collects rainwater for non-potable and potable uses. Desalination, although energy-intensive and expensive, is becoming increasingly viable in water-scarce coastal regions, offering a potential long-term solution. Rainwater harvesting, on the other hand, provides an effective, low-cost solution, especially in rural and arid regions, by capturing and storing rainwater for agricultural, industrial, and domestic uses (United Nations, 2017).

3.7.3 Integrated water resource management (IWRM)

IWRM is a holistic approach that aims to optimize the allocation and management of water resources across multiple sectors, considering the interconnectedness of water, land, and related resources. IWRM involves the active participation of various stakeholders, including government agencies, local communities, private sector actors, and environmental groups, to ensure that water resources are managed in a way that meets the needs of all users while preserving ecological sustainability. This approach emphasizes the need for balancing water supply with demand, considering environmental factors such as ecosystem health, and minimizing water-related conflicts. IWRM also promotes the efficient and equitable distribution of water resources, ensuring that vulnerable and marginalized communities have access to safe and sufficient water. By adopting IWRM, countries can address water scarcity in a more inclusive and coordinated manner, improving both the quality and quantity of available water (Global Water Partnership, 2018).

3.7.4 Climate-resilient water management (CRWM)

CRWM is increasingly important as climate change exacerbates water scarcity through altered rainfall patterns, more frequent and severe droughts, and rising temperatures. To address these challenges, strategies are needed to build resilience in both water supply and demand systems. One of the key strategies is the development of drought-resistant crops and improved agricultural practices that can thrive with limited water. For example, the use of drip irrigation, which delivers water directly to the roots of plants, reduces water wastage and increases crop yields in arid regions. Similarly, water storage systems need to be designed to withstand extreme weather events, such as heavy rainfall and floods, to ensure a stable water supply during climate-induced disruptions. Climate-resilient infrastructure also includes flood management systems, such as levees and dams, which protect communities from the effects of extreme weather while preserving water resources. Moreover, water governance frameworks must integrate climate change adaptation strategies, ensuring that policies and investments are aligned with long-term sustainability goals (IPCC, 2022).

The challenge of global water scarcity requires a multifaceted approach that integrates demand-side and supply-side management, IWRM and climate-resilient water management strategies. These approaches not only help to optimize the use and availability of water resources but also ensure that water management systems are robust enough to withstand the pressures of climate change and demographic growth. By adopting these strategies, countries and regions can work towards securing a sustainable and equitable water future for all.

3.7.5 Policy and governance: The role of strong governance in water management

Effective water governance is crucial in managing water resources and addressing the growing global water scarcity. Strong governance frameworks ensure that water resources are used sustainably, equitably distributed, and protected from over-exploitation and environmental degradation. Policy and regulatory frameworks are necessary to manage the growing demand for water in agriculture, industry, and domestic use. In the face of increasing pressures from population growth, climate change, and geopolitical tensions, countries need robust governance systems that ensure coordinated efforts across different sectors and scales. Several countries in Africa have taken steps to manage their water resources effectively, but the success of these efforts depends on the strength of governance, the implementation of policies, and the ability to engage all stakeholders in decision-making processes.

3.7.5.1 South Africa: Investment in desalination and water recycling

South Africa, particularly in urban areas like Cape Town, has experienced recurring water crises due to severe droughts and increased demand. To address these challenges, South Africa has heavily invested in desalination and water recycling as part of its water management strategy. Desalination plants, which convert seawater into potable water, have been developed along the country's coast, aiming to augment the freshwater supply for municipalities.

Additionally, the country has embraced water recycling, with many wastewater treatment plants repurposing water for industrial use, agriculture, and even potable consumption. The city of Cape Town, for example, has implemented extensive water-saving measures, including the promotion of water-efficient appliances and the recycling of water in homes and businesses (Khumalo & Mazvimavi, 2019). However, these measures are not without challenges, including the high operational costs of desalination plants and the need for continuous maintenance of infrastructure. Strong governance frameworks and collaboration between the government, private sector, and local communities are crucial to ensuring the success of these initiatives. In addition to technology-driven solutions, policies that promote water conservation and equitable access are necessary to address the root causes of water scarcity (Pereira et al., 2019).

3.7.5.2 Ethiopia: Focus on rainwater harvesting and irrigation

Ethiopia, a country highly vulnerable to water shortages, has focused on rainwater harvesting and irrigation to adapt to its water scarcity challenges. Given its reliance on rainfall for agriculture, Ethiopia has invested in large-scale rainwater harvesting projects, particularly in rural areas. These projects collect rainwater during wet seasons and store it for use during dry periods, reducing the reliance on groundwater and surface water, which are often subject to over-extraction and pollution. Additionally, Ethiopia has implemented irrigation projects to improve agricultural productivity, especially in arid regions where rainfall is inconsistent. However, challenges remain in ensuring the sustainability of these initiatives, as infrastructure maintenance and equitable distribution of water resources continue to be problematic (Fenta et al., 2021). The Ethiopian government has made strides in implementing water management policies, but there are issues related to governance and water allocation, particularly in the face of regional disparities and ongoing conflicts over water resources (Bachewe et al., 2020).

3.7.5.3 Nigeria and Kenya: Fragmented approaches due to governance challenges

In contrast to South Africa and Ethiopia, Nigeria and Kenya face more fragmented approaches to water management, primarily due to political instability, governance challenges, and corruption. Nigeria, for instance, experiences significant governance challenges, which have led to poor management of water resources. The country's water management system suffers from inadequate infrastructure, mismanagement, and inefficient policies. Additionally, disputes over water resources between states and neighboring countries—such as those involving the Niger River—further complicate governance efforts. In rural areas, groundwater resources are often over-extracted, and many communities face water contamination due to pollution and inadequate sanitation facilities (Adeyemo, 2020). The lack of a cohesive and coordinated water management policy at the national level means that many of Nigeria's water management initiatives are not effectively implemented.

Similarly, Kenya faces fragmented water governance, with water access being highly uneven across different regions. Water resources are concentrated in specific areas, while the northern parts of the country suffer from extreme water scarcity. Despite investments in water management infrastructure, including water transfer schemes and potential desalination projects, Kenya's approach to water management has been hindered by political instability, corruption, and poor implementation of policies (Ong'uti et al., 2019). Moreover, the centralization of water resources in the hands of a few large institutions and a lack of community-based management approaches have limited the effectiveness of water distribution. As in Nigeria, Kenya's water management initiatives have struggled to achieve long-term sustainability due to governance challenges, leading to a fragmented and inequitable distribution of resources.

3.7.5.4 The need for strong and inclusive water governance

To address water scarcity and ensure sustainable water management, countries need strong, transparent, and inclusive governance systems that enable the effective implementation of water policies and equitable distribution. South Africa and Ethiopia demonstrate the potential for successful water management through investments in desalination, water recycling, and rainwater harvesting. However, these solutions require effective governance and stakeholder engagement to ensure their sustainability. On the other hand, Nigeria and Kenya highlight the consequences of fragmented and weak governance, which hampers the success of water management initiatives. In both cases, political instability, corruption, and inadequate policy implementation hinder efforts to address water scarcity effectively. To overcome these challenges, countries must prioritize governance reforms, strengthen institutional capacities, and promote inclusive water management policies that involve all stakeholders, from local communities to national governments.

3.7.6 Conflict and cooperation

Water scarcity can lead to conflicts between countries, communities, and different sectors. Competition for limited water resources can escalate tensions and even trigger violence. However, cooperation and collaboration are essential to address water challenges. Sharing water resources, establishing joint management mechanisms, and promoting diplomacy can help prevent conflicts and ensure sustainable water use. International agreements, such as the Nile River Basin Initiative, provide frameworks for cooperation and equitable water distribution. By fostering dialogue, building trust, and implementing fair water-sharing arrangements, countries can mitigate water-related conflicts and promote regional stability.

Nigeria and Ethiopia are notable for political conflict over transboundary water resources (e.g., the Niger River and Nile Basin, respectively).

Countries like South Africa and Kenya have worked toward cooperation on water management, with regional agreements to share water resources more equitable distribution.

Nigeria and Ethiopia are two countries in Sub-Saharan Africa that face significant political conflict over transboundary water resources, which has led to complex diplomatic challenges and strained relations between neighboring states. The Niger River, which flows through Nigeria and several other West African countries, is one such example. The river, which is crucial for the livelihoods and economies of the countries it traverses, has been a point of contention due to competing demands for its water. Nigeria, as the largest country in the basin, has long faced tensions with its upstream neighbors, particularly in Mali and Niger, over water allocation and dam constructions (Hare, 2018). Disputes have also arisen regarding the use of water for irrigation, hydropower, and industrial purposes, which threaten the sustainable management of the river's resources. These conflicts have been exacerbated by the lack of effective regional governance structures to ensure equitable water distribution and address the needs of all countries in the basin (Niger Basin Authority, 2019).

Similarly, Ethiopia's construction of the Grand Ethiopian Renaissance Dam (GERD) on the Blue Nile has become a source of intense political conflict with downstream countries, primarily Egypt and Sudan. Egypt, which relies heavily on the Nile River for its water supply, has expressed concerns that the dam could significantly reduce its share of water, thereby exacerbating its already pressing water scarcity (Kalaydjian et al., 2021). Ethiopia argues that the dam is crucial for its development and economic growth, particularly in terms of providing reliable electricity to millions of Ethiopians. The lack of an agreement on how to manage the dam and its water usage has resulted in escalating tensions, with periodic diplomatic efforts failing to achieve a lasting resolution (Solomon, 2020). The situation highlights the complexities of transboundary water management in the context of political instability and competing national interests, underscoring the need for effective regional cooperation frameworks.

On the other hand, South Africa and Kenya have made significant strides toward regional cooperation on water management, with a focus on equitable distribution of water resources. South Africa, for instance, is a key player in the Southern African Development Community (SADC), which has developed frameworks to address water scarcity and promote collaborative water management across borders. The SADC Water Protocol, established in 2000, encourages member countries to work together on shared water resources to ensure sustainable use and prevent conflicts. South Africa shares several river basins with neighboring countries, including the Zambezi and Limpopo Rivers, and has engaged in joint efforts with Zimbabwe, Mozambique, and Botswana to manage these resources (SADC, 2020). Despite challenges such as climate change and increasing water demand, these collaborative efforts have helped to mitigate some of the risks associated with water scarcity and foster regional peace and stability.

In Kenya, cooperation on water management has been largely focused on transboundary basins such as the Tana and Ewaso Ng'iro Rivers. Kenya shares these water bodies with Ethiopia and Somalia, and over the years, the country has engaged in regional dialogues to ensure equitable distribution and address water scarcity issues. The Nairobi-based Eastern Africa Regional Environment Programme has been instrumental in promoting integrated water resource management (IWRM) in the region, encouraging countries to work together to manage shared resources efficiently (EAC, 2018). Kenya's involvement in the Nile Basin Initiative (NBI) also exemplifies its commitment to regional cooperation on water resources, despite facing challenges such as political instability and environmental degradation (Karanja et al., 2017). These efforts highlight the potential for collaboration and shared governance in managing water resources, though success is dependent on political will and effective policy implementation.

The experiences of Nigeria and Ethiopia highlight the complexities of managing transboundary water resources in politically charged environments, where national interests and competition for limited resources often lead to conflict. Conversely, the experiences of South Africa and Kenya underscore the potential benefits of regional cooperation and the importance of integrated water resource management in addressing the challenges posed by water scarcity. Effective governance, equitable distribution, and shared responsibility remain central to achieving sustainable water management solutions in these regions.

3.7.7 Conflict and cooperation over water resources in Sub-Saharan Africa

Water scarcity in Sub-Saharan Africa is not only a result of physical water shortages but also stems from complex political conflicts over the control and management of transboundary water resources. These conflicts often involve multiple countries that share vital water sources, with disputes arising over water allocation, infrastructure development, and environmental concerns. However, the region also provides examples of cooperation over water management, where countries have worked together through regional agreements to ensure equitable access to shared resources.

3.7.7.1 Conflict over transboundary water resources

Transboundary water resources, shared by multiple countries, are often sources of conflict. Differences in water needs, competing demands from agriculture, industry, and domestic use, and varying levels of economic development can exacerbate tensions. For example, the Nile River Basin, shared by ten African countries, has been a historical source of conflict due to competing claims over water resources. Mismanagement, unequal distribution, and lack of cooperation can further escalate tensions. To mitigate these challenges, international cooperation, equitable water-sharing agreements, and joint management mechanisms are crucial. By fostering dialogue, building trust, and implementing sustainable water management practices, countries can minimize conflicts and ensure the sustainable use of shared water resources.

3.7.7.2 The Niger River Basin: Political conflict in Nigeria

The Niger River is one of the most important transboundary water resources in West Africa, flowing through several countries, including Guinea, Mali, Niger, Benin, and Nigeria. However, the political context surrounding the Niger River has been marked by challenges in managing its resources equitably.

3.7.7.3 Challenges and disputes

Nigeria, as the most populous country in the basin, has historically faced tensions with its upstream neighbors, especially Mali and Niger, over water allocation. The construction of large dams, such as the Feliké Dam in Mali, has caused concern in Nigeria about reduced water flow to the Niger Delta, which is crucial for both agriculture and fisheries. Disputes over the management of water quality and the impact of upstream activities, such as mining and irrigation, have also been sources of tension.

Nigeria, as the most populous country in the Niger River Basin, faces a complex set of challenges and disputes with its upstream neighbors, particularly Mali and Niger, over the allocation and management of water resources. The Niger River is vital for Nigeria's agricultural, industrial, and domestic needs, serving as a crucial source of water for irrigation, transportation, and fishing, especially in the southern regions. With a large portion of Nigeria's population depending on the river for these essential activities, any disruption to its flow or quality can have profound socio-economic impacts. The construction of large dams, such as the Feliké Dam in Mali, has been a source of tension between Nigeria and its upstream neighbors, as these dams regulate and sometimes reduce the volume of water flowing downstream into Nigeria, particularly affecting the Niger Delta region.

The Feliké Dam, constructed on the Niger River in Mali, has raised alarms in Nigeria about the potential for reduced water availability, particularly during the dry season. This reduction in water flow is of significant concern to Nigeria, as the Niger Delta is a critical agricultural and fishing zone, where local communities rely heavily on the river for irrigation and fish production. The reduction in water flow could lead to decreased agricultural productivity, loss of fisheries, and negatively impact the livelihoods of millions of Nigerians. This situation is particularly alarming as the Niger Delta is not only an essential source of food but also a vital ecological zone supporting rich biodiversity. Therefore, any alterations in water flow and volume due to upstream developments can result in both immediate and long-term economic consequences for downstream countries like Nigeria (UNEP, 2017).

Another source of tension between Nigeria and its upstream neighbors is the impact of upstream activities such as mining, irrigation, and industrial use, which degrade the quality of the water flowing downstream. In countries like Mali and Niger, large-scale irrigation projects and mining operations often result in the contamination of the river with harmful chemicals, sedimentation, and pollutants. This deteriorates the water quality, making it unfit for consumption and damaging the ecosystems downstream in Nigeria. The pollution caused by mining, which releases harmful substances like heavy metals and toxic chemicals into the water, has serious consequences for public health and the environment. Communities in Nigeria, particularly those living along the river, are at risk of exposure to these pollutants, which can lead to waterborne diseases, agricultural losses, and long-term health problems (Falkenmark & Rockström, 2010).

In addition to the concerns over water flow and quality, the governance of transboundary water resources in the Niger River Basin has been marked by a lack of cooperation and effective management. Despite the river being shared by multiple countries, including Mali, Niger, and Nigeria, there has been no comprehensive and enforceable agreement on water allocation and usage. This absence of a unified framework for managing the river's resources leads to disagreements on priorities and conflicts over how water should be distributed, especially during periods of scarcity. The absence of clear rules and mechanisms for resolving conflicts exacerbates tensions, making it difficult to address shared water challenges collectively. As a result, each country seeks to maximize its own access to water resources, often at the expense of others, which can lead to increased political instability and disputes (Rogers, 2006).

In the face of these challenges, Nigeria has called for greater regional cooperation and the establishment of binding agreements on the equitable distribution of water resources in the Niger River Basin. Effective water management in the basin would require transparent communication, cooperation, and joint efforts to tackle the shared issues of water allocation, water quality, and environmental degradation. The establishment of a regional framework, such as the Niger Basin Authority (NBA), has been a step toward cooperation, but the institution faces challenges in enforcing agreements and ensuring compliance from all parties involved. Until comprehensive and effective governance mechanisms are in place, the Niger River Basin will continue to be a site of political tensions and disputes, with water scarcity and quality issues posing serious threats to the livelihood and development of millions of people in the region (Sadoff & Grey, 2002).

3.7.8 Governance and cooperation

Despite these tensions, the countries that share the Niger River Basin have taken steps toward regional cooperation. The Niger Basin Authority (NBA), established in 1980, works to manage the water resources of the Niger River, promote sustainable development, and resolve conflicts. However, the effectiveness of the NBA has often been challenged by political disagreements, lack of infrastructure, and national priorities that overshadow collective action (Salami et al., 2016).

The Niger River Basin, shared by several countries including Mali, Niger, and Nigeria, has witnessed both cooperation and tension in the management of its water resources. Despite the ongoing disputes, the establishment of the Niger Basin Authority (NBA) in 1980 marked an important step toward regional cooperation. The NBA was created with the objective of fostering sustainable management of the Niger River's water resources, promoting the economic development of the basin countries, and resolving conflicts related to water use. The authority aims to ensure that the river's resources are equitably shared and utilized for the benefit of all the countries that depend on it. However, while the NBA has been instrumental in providing a platform for dialogue, its effectiveness has often been hindered by political disagreements among member states, as well as the lack of adequate infrastructure to support coordinated water management efforts (Salami et al., 2016).

One of the primary challenges facing the NBA is the political instability and competing national interests that often take precedence over regional cooperation. Each country in the basin has its own priorities regarding water use, which complicates the process of reaching collective agreements. For example, upstream countries like Mali and Niger may prioritize the construction of dams and irrigation systems to support their agricultural sectors, while downstream countries like Nigeria are concerned about the impact of these developments on water availability and quality. These conflicting interests often lead to disagreements over how water resources should be allocated, particularly during times of scarcity. Moreover, the NBA's ability to enforce agreements and ensure compliance among member states remains limited, which undermines its overall effectiveness in managing the shared water resources of the Niger River (Giordano et al., 2003).

In addition to political tensions, the NBA faces significant challenges related to the lack of infrastructure and technical capacity to support effective water management. Many of the countries in the Niger River Basin lack the financial resources and technical expertise needed to implement large-scale water management projects or to monitor the river's flow and quality consistently. This has hindered the NBA's ability to carry out its mandate effectively. Furthermore, the river's vast and complex hydrology, which spans over several countries, makes it difficult to manage in a coordinated and integrated manner. Without adequate infrastructure for water storage, treatment, and distribution, and without a unified approach to monitoring and assessing water quality, the NBA's efforts to promote sustainable water use and prevent conflicts are often thwarted (Salami et al., 2016).

Another significant barrier to effective governance in the Niger River Basin is the prioritization of national over regional interests. While the NBA promotes regional cooperation, many of the member states continue to prioritize their own development goals, which may not always align with the shared objectives of the basin. For instance, large-scale irrigation projects in upstream countries may benefit their local economies but could reduce the amount of water flowing downstream, negatively impacting agriculture and livelihoods in countries like Nigeria. Additionally, political instability and governance challenges in certain basin countries complicate the ability of the NBA to carry out its initiatives effectively. Without a strong commitment to collective action and the resolution of conflicts, the potential for collaboration within the Niger River Basin remains limited (Vermuelen & Garschagen, 2014).

Despite these challenges, there have been some positive developments in terms of regional cooperation in the basin. The NBA has facilitated dialogue among member states, leading to agreements on certain water management issues, such as the establishment of joint monitoring systems and the promotion of sustainable irrigation practices. Additionally, the NBA has sought to raise awareness of the need for transboundary cooperation, with the goal of ensuring that the Niger River remains a source of sustainable development for all basin countries. However, for the NBA to be truly effective, there is a need for stronger political will, improved infrastructure, and a greater focus on shared benefits rather than national interests. Strengthening the governance framework and ensuring that all basin countries are actively engaged in the decision-making process will be essential for overcoming the existing challenges and promoting long-term cooperation in the management of the Niger River's water resources (Salami et al., 2016).

3.7.9 The Nile River Basin: Disputes over water allocation

The Nile River is another key transboundary water resource in Africa, flowing through eleven countries, with Ethiopia, Egypt, and Sudan being the most prominent stakeholders. The political disputes over the Nile Basin are long-standing, primarily revolving around water allocation and control.

3.7.9.1 The grand Ethiopian renaissance dam (GERD)

The construction of Ethiopia's Grand Ethiopian Renaissance Dam (GERD) on the Blue Nile has been a major source of tension between Ethiopia, Egypt, and Sudan. Egypt has expressed concerns that the GERD will reduce its share of the Nile's water, which is crucial for its agriculture and drinking water supply, given that Egypt is already one of the world's most water-scarce countries. Ethiopia, on the other hand, views the dam as essential for its energy generation and economic development.

The Grand Ethiopian Renaissance Dam (GERD) has become a central point of contention in the Nile Basin, particularly between Ethiopia, Egypt, and Sudan. The dam, located on the Blue Nile in Ethiopia, is Africa's largest hydroelectric power project and is seen by Ethiopia as a crucial infrastructure project for its economic development. The country views the dam as a means of addressing its energy deficit, fostering industrialization, and driving economic growth. With the potential to generate over 6,000 megawatts of electricity, the GERD could provide much-needed power to millions of Ethiopians and significantly improve the country's energy infrastructure (Kefale, 2020).

However, the construction of the GERD has raised significant concerns among Egypt and Sudan, particularly regarding the potential impact on their water security. Egypt, which relies on the Nile River for nearly all of its freshwater supply, has voiced fears that the dam will reduce the amount of water flowing downstream, which is critical for its agriculture, drinking water, and overall survival. Egypt is already one of the world's most water-scarce countries, with the Nile accounting for over 95% of its water resources. Consequently, any reduction in the flow of the river could exacerbate Egypt's already dire water scarcity issues (Ali et al., 2019). The fear is that the filling of the GERD's reservoir could cause a significant reduction in the water available to Egypt, especially during periods of drought or in the early stages of the dam's operation.

Sudan also has concerns, although its position is more nuanced. While Sudan stands to benefit from the GERD's electricity generation, it has expressed worries about the dam's effect on its own water management and the operation of its dams along the Nile. Sudan's agricultural sector is also highly dependent on the Nile, and disruptions to water flow or changes in the river's flow regime could affect irrigation practices and water availability. Additionally, the dam's construction has raised concerns about the safety of Sudanese dams located downstream, with the potential for flooding or damage during the filling of the GERD's reservoir (Hassan, 2021).

The dispute over the GERD highlights the complex nature of transboundary water management in the Nile Basin, where multiple countries share the river's resources. The Nile River is one of the world's longest rivers, flowing through 11 countries, and it has historically been governed by agreements that heavily favored Egypt, particularly the 1959 Nile Waters Agreement between Egypt and Sudan. This agreement, which allocated 90% of the river's flow to Egypt and Sudan, did not include other upstream countries like Ethiopia, which has since led to growing tensions over water allocation as these upstream countries seek to exert greater control over the Nile's resources (Elhance, 2000).

Ethiopia, in particular, has argued that it has the right to develop its water resources for the benefit of its population, emphasizing that the GERD is a necessary step for addressing the country's energy needs and lifting millions of people out of poverty. From Ethiopia's perspective, the dam is an essential part of its development strategy and has the potential to significantly enhance the country's economy by attracting investment and boosting industrialization. The country has also made it clear that it views the GERD as an issue of national sovereignty and development, which has contributed to its reluctance to reach a binding agreement with Egypt and Sudan on the operation and filling of the dam's reservoir (Kefale, 2020).

The tension over the GERD has led to a series of diplomatic negotiations, with the African Union, the United Nations, and other international bodies attempting to mediate between the involved parties. However, despite various rounds of negotiations, a final agreement on the dam's filling and operation has yet to be reached, with the dispute often spilling into the public domain. The situation remains precarious, with the potential for further escalation if the concerns of downstream countries are not addressed through equitable agreements and cooperative management frameworks.

In conclusion, the GERD dispute underscores the challenges of managing shared water resources in a region with growing populations, changing climate patterns, and economic pressures. It highlights the need for robust, inclusive, and cooperative water governance mechanisms that can address the interests of all stakeholders while promoting sustainable development. The resolution of the GERD issue will require a delicate balance between national development priorities and the collective good of all Nile Basin countries.

3.7.9.2 Diplomatic tensions

The dispute has escalated over the filling and operation of the GERD, with repeated rounds of talks under the auspices of the African Union and the United Nations failing to yield a comprehensive agreement. Although the conflict has not escalated into armed confrontation, the potential for future conflict remains high, particularly as climate change may further strain the Nile's water availability.

The diplomatic tensions surrounding the Grand Ethiopian Renaissance Dam (GERD) have grown over the years, with the filling and operation of the dam becoming a contentious issue between Ethiopia, Egypt, and Sudan. Egypt has viewed Ethiopia's construction of the GERD on the Blue Nile as a direct threat to its historical water rights to the Nile, as the country relies heavily on the river for both agricultural irrigation and drinking water. Egypt fears that filling the GERD's reservoir will drastically reduce the water flow downstream, jeopardizing its already scarce water resources. Sudan, while acknowledging the potential benefits of the dam, particularly in terms of electricity generation, shares concerns about the impact on its water infrastructure and agriculture due to the dam's operation and filling process (Elhance, 2000; Kefale, 2020; Salami & Opeyemi, 2016).

Repeated rounds of negotiations have taken place under the auspices of the African Union (AU), the United Nations (UN), and other international organizations, but none have resulted in a comprehensive agreement. The negotiations have focused on issues such as the speed at which the reservoir is filled, the regulation of water flow, and how to manage potential droughts or other emergency scenarios. Despite the involvement of various mediators, the negotiations have failed to reach a consensus, primarily due to differing national priorities. Ethiopia, which sees the GERD as essential for its energy generation and economic development, is keen to proceed with the filling of the dam to meet its growing energy demands. On the other hand, Egypt, which already suffers from extreme water scarcity, views the dam as a critical issue for its survival and agricultural productivity, thereby demanding a legally binding agreement to safeguard its water rights.

One of the primary sticking points has been the lack of a legally binding agreement that would regulate the dam's operation and ensure the equitable distribution of water among the three countries. Egypt insists on a comprehensive deal that would guarantee a consistent and reliable flow of water, while Ethiopia argues that it has the right to develop its resources for its own national benefit, particularly in light of the country's severe electricity shortages. Sudan, meanwhile, has taken a more neutral stance, expressing concerns over the dam's potential impact on its own infrastructure, but also recognizing the potential for regional benefits, such as increased energy supply and flood control. However, Sudan has also sought guarantees that its own water needs will not be adversely affected.

The failure to reach a comprehensive agreement, despite several rounds of negotiations, has raised concerns about the potential for conflict. While the situation has not yet escalated into armed confrontation, tensions remain high, and the possibility of future conflict is a real concern. The dispute over the GERD highlights the growing importance of water security in the region, particularly as climate change begins to exacerbate water scarcity. The Nile, already under stress due to increasing populations and water demand, could become even more unpredictable due to climate change. Increased evaporation, fluctuating rainfall patterns, and prolonged droughts could further reduce the river's water availability, intensifying competition among the countries that rely on the Nile for their water supply.

Moreover, the lack of an agreement on the GERD could set a dangerous precedent for future conflicts over shared water resources. Transboundary water disputes, such as this one, are likely to become more common in a world where climate change accelerates the need for effective water governance. The Nile Basin, home to over 300 million people, is one of the most complex transboundary water systems in the world, with multiple countries sharing its water resources. Without a comprehensive and binding agreement, the potential for tensions to escalate into broader regional conflict remains high.

As climate change continues to strain water resources across the globe, the GERD dispute serves as a stark reminder of the delicate balance required to manage shared water resources. International cooperation, transparent communication, and a willingness to compromise will be essential in resolving this conflict and ensuring that the Nile Basin countries can meet their water needs without compromising regional stability. The importance of adopting a cooperative framework for water management, one that considers the needs of all stakeholders and prioritizes sustainability, cannot be overstated. It is clear that only through diplomacy, collaboration, and fair negotiation can the GERD issue be resolved in a way that benefits all parties involved.

3.8 Cooperation in Transboundary Water Management

While conflict is prevalent, there are also notable examples of cooperation among countries sharing water resources. Cooperation has often been driven by the recognition that shared water bodies are critical to the region's agricultural productivity, energy generation, and economic development.

3.8.1 The Zambezi River Basin: Cooperation between Countries

Several countries, including Angola, Namibia, Zambia, Zimbabwe, Mozambique, and Tanzania share the Zambezi River Basin. The river's water resources have historically been subject to tensions, particularly between Zimbabwe and Zambia, but over time, cooperative frameworks have been established.

3.8.2 The Zambezi River Authority (ZRA)

The Zambezi River Authority was established in 1987 to promote the joint management of the river's water resources, particularly for the operation of the Kariba Dam and the Cahora Bassa Dam. The ZRA has been successful in facilitating cooperation in areas such as energy generation, flood control, and the management of water for agriculture. The Zambezi Watercourse Commission (ZAMCOM), established in 2004, focuses on improving the collaborative management of water resources in the entire basin, ensuring that each member country's interests are considered.

The Zambezi River Authority (ZRA) was established in 1987 with the primary objective of managing the water resources of the Zambezi River, specifically focusing on the operation of two significant dams: the Kariba Dam and the Cahora Bassa Dam. The ZRA was created as a joint body between Zambia and Zimbabwe, two of the countries sharing the Zambezi River Basin. The cooperation between these two nations was vital for the operational management of these dams, which are crucial for electricity generation and water storage, particularly for the region's agricultural and industrial needs (Zambezi River Authority, 2020).

One of the key successes of the ZRA has been in the area of energy generation. Both the Kariba and Cahora Bassa Dams are critical sources of hydroelectric power for Zambia, Zimbabwe, and other countries in the Southern African region. The authority has facilitated the joint operation of these dams, ensuring the equitable distribution of the electricity generated. Additionally, the ZRA has been instrumental in flood control, mitigating the risks of devastating floods by coordinating the release of water from the dams. This cooperation has led to improved stability in the river system, reducing the impacts of seasonal flooding on local communities and agriculture (Sharma & Nkamleu, 2017).

In 2004, the Zambezi Watercourse Commission (ZAMCOM) was established to broaden the scope of regional cooperation. While the ZRA is specifically focused on the management of the two dams, ZAMCOM aims to ensure collaborative management of water resources across the entire Zambezi River Basin. This broader mandate includes all 8 riparian countries: Zambia, Zimbabwe, Mozambique, Angola, Namibia, Malawi, Tanzania, and Botswana. ZAMCOM's role is to facilitate dialogue, cooperation, and coordination among member states, ensuring that the interests of all countries are considered in decisions regarding water usage, conservation, and development. ZAMCOM also works on implementing a shared vision for sustainable water resource management, addressing issues such as pollution, water scarcity, and the effects of climate change on the water availability in the basin (Zambezi Watercourse Commission, 2020).

ZAMCOM's collaborative approach is particularly significant as the Zambezi River Basin faces increasing pressure from climate change, population growth, and industrialization. The commission aims to achieve equitable distribution of water resources, balancing the needs for agriculture, energy generation, and domestic use. This integrated management approach helps to prevent conflicts over water allocation and ensures that the resources of the basin are used sustainably for future generations (Mwandira et al., 2019).

However, despite these successes, challenges remain. There are ongoing concerns about the equitable sharing of water resources, particularly among countries with competing needs. For example, the construction of dams upstream can affect the availability of water downstream, potentially leading to disputes. Furthermore, climate change poses a growing threat to the stability of the basin's water resources, making it more difficult to predict water flows and necessitating adaptive management strategies (Mutasa et al., 2018).

The Zambezi River Authority and ZAMCOM represent important efforts in the cooperative management of the Zambezi River Basin. Through their joint efforts, these institutions have contributed to improved regional cooperation, energy production, flood management, and agricultural sustainability. However, as the basin faces new challenges, especially from climate change, ongoing cooperation and adaptive management will be essential to ensure the continued equitable and sustainable use of Zambezi's water resources.

3.8.3 Transboundary water governance

Despite challenges, countries in the Zambezi Basin have largely agreed on the principles of equitable and reasonable utilization of water, ensuring that no single country can unilaterally exploit the river's resources. The regional cooperation in the Zambezi Basin is considered a success in promoting peace and sustainable water resource management (Ochieng, 2017).

Transboundary water governance in the Zambezi Basin represents a model of cooperation despite the many challenges associated with managing shared water resources. The principles of equitable and reasonable utilization, enshrined in international water law, have been fundamental in guiding the cooperative efforts between the riparian countries. These principles ensure that no single country can exploit the river's resources in a way that undermines the rights or needs of others, fostering an environment of mutual respect and collaboration. This framework aims to balance the diverse needs of the countries that share the Zambezi River, which include agriculture, hydropower generation, domestic consumption, and industrial development (Ochieng, 2017).

The establishment of the Zambezi River Authority (ZRA) and the Zambezi Watercourse Commission (ZAMCOM) has been pivotal in promoting regional cooperation. These institutions play key roles in coordinating efforts between the riparian countries to ensure that water management decisions are made transparently and inclusively. ZRA's work, especially in the operation of the Kariba and Cahora Bassa dams, has been crucial in facilitating energy production, flood control, and irrigation, benefiting not only the countries directly involved—Zambia and Zimbabwe—but also other nations in the region. ZAMCOM, on the other hand, takes a more holistic approach, focusing on the long-term, sustainable management of the entire Zambezi River Basin, including its tributaries and surrounding ecosystems (Zambezi River Authority, 2020).

The success of the regional cooperation can be seen in the way these institutions have navigated complex issues, such as water sharing and the construction of infrastructure. For example, the Kariba Dam, which is jointly managed by Zambia and Zimbabwe, has played a central role in the region's energy production. The ZRA's ability to maintain a balance between the needs of both countries and ensure that water releases from the dam are done equitably has helped prevent conflicts over water usage. Similarly, the establishment of ZAMCOM in 2004 broadened the scope of cooperation, bringing together all eight riparian countries in a more inclusive governance framework that emphasizes the shared responsibility for the basin's resources (Sharma & Nkamleu, 2017).

Despite the progress made, the Zambezi Basin still faces challenges. Climate change, population growth, and increasing industrial demands are putting pressure on water resources, and countries must continue to work together to address these issues. For instance, downstream countries, such as Mozambique, are highly dependent on the water flowing from upstream dams, making them vulnerable to the decisions made by their upstream neighbors. Additionally, the construction of new dams or infrastructure projects can create tensions if not carefully coordinated with other countries sharing the basin. Nevertheless, the ongoing dialogue facilitated by ZAMCOM and the ZRA has helped manage these tensions and avoid major conflicts, showcasing the potential of cooperative water governance (Mwandira et al., 2019).

In conclusion, the governance of the Zambezi River Basin is a testament to the benefits of regional cooperation in managing shared water resources. The principles of equitable and reasonable utilization have served as the foundation for collaborative efforts, ensuring that all riparian countries have a voice in decision-making processes. Despite the challenges posed by climate change and increasing demands on water resources, the Zambezi River Authority and Zambezi Watercourse Commission continue to provide a framework for sustainable and peaceful management of the basin's water, making it a successful model of transboundary water governance (Ochieng, 2017).

3.9 The Orange-Senqu River Basin: Joint Management Between South Africa and Lesotho

The Orange-Senqu River flows through Lesotho, South Africa, and Namibia and has long been an area of regional cooperation for water management. The

Orange-Senqu River Commission (ORASECOM) was established in 2000 to promote cooperative management of the river's resources, focusing on equitable water distribution, especially for irrigation and domestic use.

3.9.1 Integrated water resource management

ORASECOM's approach has been relatively successful in managing the river's resources, particularly through the shared management of the Lesotho Highlands Water Project (LHWP). This large-scale water transfer project supplies water from Lesotho to South Africa's industrial heartland. The LHWP has provided a win-win solution to both countries: Lesotho benefits from the project's hydropower generation, while South Africa receives a stable water supply.

Integrated Water Resource Management (IWRM) in the Orange-Senqu River Basin, which is managed by the Orange-Senqu River Commission (ORASECOM), has been a relatively successful model of cooperation, particularly through the shared management of the Lesotho Highlands Water Project (LHWP). The LHWP is a large-scale water transfer project that supplies water from Lesotho to South Africa's industrial heartland, addressing critical water supply and energy needs for both countries. This project exemplifies how countries sharing a water resource can collaborate to develop mutually beneficial solutions, balancing the interests of upstream and downstream nations in a way that fosters long-term sustainability.

The LHWP serves as a win-win solution for both Lesotho and South Africa. For Lesotho, the project not only provides a reliable source of hydropower generation but also contributes significantly to the country's economy. The hydropower plants built as part of the LHWP have become a key element in Lesotho's energy infrastructure, supporting domestic energy needs and creating potential for export revenue. In addition, the financial benefits from the water transfer fees paid by South Africa help Lesotho to fund its national development projects, making it a valuable source of income (Cairns & O'Keeffe, 2018).

For South Africa, the LHWP provides a stable and reliable water supply to the arid regions of the country, particularly the industrial heartland in Gauteng, which faces chronic water scarcity. The project ensures a consistent flow of water that is critical for agricultural production, industrial processes, and domestic consumption. It has thus become an essential part of South Africa's water security strategy, contributing to its economic stability and growth. Without this project, South Africa would face increased vulnerability to water shortages, especially given the challenges posed by climate change and increasing population pressure (Le Roux, 2019).

The success of the LHWP highlights the importance of IWRM in transboundary water governance. It shows that when countries share water resources, cooperative management can yield economic, environmental, and social benefits for all parties involved. The project is managed through a set of agreements and protocols that ensure the equitable distribution of water, the efficient operation of the water transfer systems, and the shared responsibilities of both countries in maintaining the infrastructure (Zaranyika & Makurira, 2020). The cooperation between Lesotho and South Africa under the framework of ORASECOM is a prime example of how IWRM can be applied to manage complex water systems in a way that supports both regional development and environmental sustainability.

Despite its success, the LHWP also faces challenges, particularly related to environmental concerns and the long-term sustainability of water resources. There have been ongoing debates about the environmental impacts of the water transfer, such as the effects on the downstream river ecosystems and the potential for water scarcity in the future due to climate change. However, ORASECOM's holistic approach to IWRM has helped to address these issues by integrating environmental considerations into the decision-making process, promoting sustainable water management practices, and encouraging the responsible use of water resources (Schreiner & Van Koppen, 2015).

The Lesotho Highlands Water Project demonstrates the potential for successful transboundary water management when countries work together under a framework like ORASECOM. By sharing the benefits of water resources and ensuring that both nations are able to meet their development goals, the LHWP provides a model for future water resource management projects in other parts of the world. The project's success underscores the importance of integrated, cooperative, and sustainable water management practices in addressing the growing challenges of water scarcity and environmental degradation in transboundary river basins.

3.9.2 Challenges

Despite these successes, climate variability and demand growth are likely to put pressure on the basin's resources in the future, and cooperation will need to be strengthened to address these emerging challenges (Turton, 2014).

Despite the significant successes of the Lesotho Highlands Water Project (LHWP) and the broader framework of Integrated Water Resource Management (IWRM) in the Orange-Senqu River Basin, several challenges threaten these efforts' sustainability. One of the primary concerns is climate variability, which has already begun to impact the region. The basin is susceptible to both droughts and floods, with changing precipitation patterns and rising temperatures expected to exacerbate these risks in the future (Turton, 2014). These climate-induced changes could lead to decreased water availability in some years, especially given the already water-scarce conditions in South Africa and the reliance on water transfers from Lesotho. Droughts, in particular, could limit the flow of water through the LHWP, potentially leading to water shortages that disrupt agriculture, industrial processes, and domestic water supply.

In addition to climate variability, population growth and economic development are expected to increase the demand for water resources in the basin. South Africa, in particular, faces significant challenges related to water demand, as the country's population continues to grow and urbanization expands. The increasing need for water for agriculture, industry, and households could strain the already limited water resources, particularly in the Gauteng region, which relies heavily on water transfers from Lesotho. The growing demand for water will require careful management to ensure equitable distribution among the various sectors and users, as well as between Lesotho and South Africa (Le Roux, 2019).

The combined pressures of climate change and rising water demand will likely require enhanced cooperation among the countries sharing the basin. While the Orange-Senqu River Commission (ORASECOM) has played a vital role in fostering collaboration and managing water resources, future challenges may demand even stronger institutional frameworks and more robust governance mechanisms. Strengthening cooperation will be essential to address the emerging challenges of water scarcity, pollution, and ecosystem degradation. Furthermore, continued investments in water-saving technologies, improved water infrastructure, and more efficient agricultural practices will be necessary to ensure that the basin's resources are managed sustainably in the face of these growing pressures.

As the region confronts these challenges, it is crucial for ORASECOM and the relevant stakeholders to prioritize long-term planning and adaptive management strategies. These strategies must account for the uncertainties associated with climate change and future water demands, and they must also involve comprehensive risk assessments and contingency plans. In doing so, the countries in the Orange-Senqu River Basin can work towards ensuring the sustainability of the LHWP and other collaborative water management initiatives, ultimately securing water resources for future generations (Schreiner & Van Koppen, 2015).

In Sub-Saharan Africa, water scarcity and management are deeply intertwined with political tensions and efforts at cooperation. Conflicts over water resources, such as those in the Niger and Nile River Basins, reflect the region's challenges in balancing national interests with regional needs. On the other hand, cooperative efforts in the Zambezi, Orange-Senqu, and other river basins highlight the potential for collaboration in addressing water scarcity and promoting sustainable development. Moving forward, it will be crucial for African countries to strengthen these cooperative frameworks to address the growing challenges posed by water scarcity, climate change, and population growth.

3.10 Impact of Water Scarcity in Sub-Saharan Africa: Reliance on Desalination, Conflict Over Water Resources, and Severe Agricultural Limitations

Sub-Saharan Africa (SSA) is facing significant water scarcity challenges due to a combination of climate change, population growth, and socio-economic factors. Water scarcity in this region impacts various sectors, including agriculture, industry, and the daily lives of people. In response to these challenges, nations have had to rely on alternative solutions such as desalination, confront conflicts over shared water resources, and cope with severe agricultural limitations. This article explores the three key impacts of water scarcity in the region: reliance on desalination, conflict over water resources, and severe agricultural limitations.

3.10.1 Reliance on desalination

Desalination, the process of converting seawater into freshwater, has been employed by a few coastal nations in Sub-Saharan Africa to address the challenge of water scarcity. However, the use of desalinated water is still limited in the region due to high costs and technological constraints.

3.10.1.1 Limited implementation

Unlike the MENA region, where desalination is widespread, Sub-Saharan Africa has not heavily invested in desalination technologies due to high operational costs and energy demands. For instance, countries like South Africa and Namibia have initiated desalination projects, but these remain relatively small-scale. As of 2020, desalination capacity in Sub-Saharan Africa was limited to around 2.6 million cubic meters per day, a tiny fraction of the region's total water demand (UNEP, 2020).

Desalination has the potential to provide a critical source of fresh water for countries in water-scarce regions, but its implementation in Sub-Saharan Africa remains limited. The Middle East and North Africa (MENA) region has long been a leader in desalination, with large-scale facilities providing a significant portion of their fresh water. In contrast, Sub-Saharan Africa has not heavily invested in desalination technologies due to a range of challenges, primarily high operational costs, energy demands, and financial constraints. Although countries like South Africa and Namibia have initiated desalination projects, these efforts are relatively small-scale compared to global standards. As of 2020, desalination capacity in Sub-Saharan Africa stood at only about 2.6 million cubic meters per day, a fraction of the region's total water demand (UNEP, 2020). This limited capacity highlights the barriers the region faces in utilizing desalination as a widespread solution to water scarcity.

One of the major barriers to scaling up desalination in Sub-Saharan Africa is the high operational cost, particularly the energy requirements. Desalination is an energy-intensive process, and many Sub-Saharan African countries struggle with unreliable electricity supplies and high energy costs. The cost of desalinated water is often significantly higher than water from traditional sources, such as surface or groundwater, making it unaffordable for many communities (Fischer & Salameh, 2019). While renewable energy sources such as solar power offer some promise for reducing operational costs, the infrastructure needed to harness these resources remains underdeveloped in many areas. Additionally, the installation and maintenance of desalination plants require significant financial investments, which are often beyond the reach of many Sub-Saharan countries, especially those that are already struggling with poverty and limited access to capital.

In countries where desalination projects have been implemented, such as South Africa and Namibia, these projects have provided essential water supplies, particularly in coastal urban areas where freshwater resources are limited. In Cape Town, South Africa, the city faced a severe drought in the late 2010s, which highlighted the potential of desalination as a supplementary water source. The city implemented a small-scale desalination plant to supplement its water supply during the crisis, though it has since faced criticism for the plant's high operational costs and limited impact on long-term water security (Hoffman, 2019). Similarly, Namibia has invested in desalination to address water shortages in its coastal cities, but the high costs have limited the scalability of such projects across the country.

Despite these challenges, desalination remains a potential solution to water scarcity in Sub-Saharan Africa, particularly in areas where freshwater resources are inadequate or unreliable. However, to make desalination more viable and sustainable in the region, governments and international partners will need to invest in infrastructure, renewable energy solutions, and financing mechanisms that can reduce costs and increase accessibility. Additionally, desalination should be considered as part of a broader integrated water resource management strategy that includes water conservation, wastewater recycling, and improved water distribution networks to address both the supply and demand sides of water management.

3.10.1.2 Energy constraints

Desalination requires substantial amounts of energy, which poses a significant challenge for many Sub-Saharan African countries that face **energy deficits** and depend heavily on fossil fuels or unreliable renewable sources for power. For instance, Namibia's desalination plants rely on power generated from coal, which raises concerns about sustainability and environmental impact (World Bank, 2019).

Energy constraints present a significant challenge to the widespread adoption of desalination in Sub-Saharan Africa, particularly because desalination is an energy-intensive process. Desalination technologies, such as reverse osmosis, require considerable electricity to separate salt from seawater, and this high energy demand makes it difficult for many countries in the region to sustain large-scale desalination projects. Many Sub-Saharan African nations already face significant energy deficits, relying heavily on fossil fuels such as coal, oil, or natural gas, as well as intermittent and unreliable renewable energy sources like solar or wind (Kadi et al., 2021). This dependency on fossil fuels not only makes desalinated water more expensive but also raises environmental concerns about the sustainability of such operations.

For example, Namibia's desalination plants, which serve as a critical source of water for its coastal cities, primarily rely on power generated from coal-fired power stations (World Bank, 2019). While coal remains an affordable energy source in the short term, it contributes to environmental degradation through carbon emissions, making the desalination process less sustainable in the long run. Additionally, reliance on coal or other fossil fuels for desalination contradicts global goals of reducing greenhouse gas emissions and addressing climate change. In the context of Sub-Saharan Africa, where access to clean energy is already limited, adding desalination as a major energy consumer places further strain on the region's power grid.

The use of non-renewable energy for desalination also complicates the long-term viability of these projects. As energy prices fluctuate and the push for cleaner energy solutions increases, the financial sustainability of desalination becomes uncertain if energy costs remain high. The shift toward renewable energy sources, such as solar and wind, offers a potential solution to reduce the carbon footprint of desalination. However, the infrastructure to harness and store renewable energy on a large scale is underdeveloped in many Sub-Saharan countries, and the initial investment required for such technologies is often prohibitive (Boccaletti et al., 2020).

To address these challenges, some countries are exploring hybrid systems that combine renewable energy sources with desalination plants to reduce dependence on fossil fuels. For example, the Desalination Project in Cape Verde, which uses a combination of solar energy and wind power to help reduce energy costs, shows promise for scaling up renewable-powered desalination in the region (Boccaletti et al., 2020). Similarly, the development of small-scale, off-grid desalination systems powered by solar energy could offer a more sustainable and decentralized approach to providing clean water to remote communities.

Ultimately, overcoming energy constraints will be essential for the future of desalination in Sub-Saharan Africa. Investments in renewable energy infrastructure, as well as improvements in energy efficiency, will be crucial to make desalination both economically viable and environmentally sustainable. Furthermore, integrating desalination into broader water management strategies—such as improving water conservation, expanding wastewater treatment and recycling, and enhancing water storage and distribution systems—will help ensure that the region can address its growing water needs without exacerbating energy deficits or environmental harm.

3.10.1.3 Environmental impact

Like in other regions, desalination plants in Sub-Saharan Africa produce brine waste, which is often released back into the sea, potentially harming marine ecosystems. While this impact has been studied in areas like the Cape Town desalination plants, there are concerns about scaling up desalination without addressing environmental risks (UNESCO, 2019).

Environmental impact is a significant concern when it comes to desalination in Sub-Saharan Africa, particularly with respect to the disposal of brine, the byproduct of the desalination process. Brine is a highly concentrated solution of salt, chemicals, and minerals that is typically discharged back into the sea or other bodies of water. When released in high volumes, brine can lead to an increase in salinity, which can disrupt marine ecosystems and negatively affect aquatic life. This issue is especially relevant in Sub-Saharan Africa, where desalination projects are often concentrated in coastal areas, such as in Cape Town, South Africa, and in coastal cities in Namibia (UNESCO, 2019).

The environmental risks associated with brine disposal have been well-documented, particularly in regions like the Mediterranean and the Middle East, where desalination has been more widespread. However, as desalination becomes more common in Sub-Saharan Africa, concerns are growing about the potential long-term effects of brine on marine biodiversity. Brine discharge can alter the physical and chemical properties of seawater, leading to local changes in temperature, salinity, and oxygen levels. These changes can make it difficult for marine organisms to survive, potentially disrupting fisheries that many coastal communities rely on for food and livelihood. Studies conducted on desalination plants in Cape Town have shown that brine can cause localized mortality of marine organisms and may also affect the reproductive cycles of fish species (UNESCO, 2019).

In addition to the brine itself, desalination plants often require the use of chemicals for cleaning and maintenance, such as anti-scaling agents and biocides. These chemicals can further harm marine life when they are discharged into the ocean alongside brine, adding to the environmental burden (UNESCO, 2019). Furthermore, the energy consumption required for desalination, which is often derived from fossil fuels, contributes to air pollution and greenhouse gas emissions, exacerbating climate change and its associated environmental impacts.

The challenge of scaling up desalination without addressing these environmental risks is a key issue for Sub-Saharan Africa, where many countries are already grappling with the dual threats of water scarcity and environmental degradation. Although desalination presents a potential solution to water shortages, particularly in arid regions, its environmental costs cannot be ignored. Effective management of brine disposal, along with the development of cleaner, more sustainable desalination technologies, will be crucial to mitigate the adverse effects on marine ecosystems. Some countries in the region, such as South Africa, have been exploring solutions such as deep-water disposal or the use of brine-mineral recovery technologies, which could help to reduce the impact on the marine environment (UNESCO, 2019).

Furthermore, integrating desalination into a broader, more sustainable water management strategy could help alleviate some of the environmental pressures. For example, combining desalination with water conservation efforts, wastewater recycling, and improved water infrastructure could reduce the overall demand for desalinated water and minimize the environmental footprint of desalination projects. It is also essential to continue conducting environmental impact assessments and developing best practices for desalination projects to ensure that they do not inadvertently harm the ecosystems they are meant to serve.

3.10.1.4 Conflict over water resources

Water scarcity in Sub-Saharan Africa often intersects with political tensions, leading to conflict over shared water resources. Many countries in the region share transboundary rivers and lakes, which increases the potential for disputes over access and control.

3.10.1.5 Key conflict areas

Nile River Basin: The Nile is shared by 11 countries, including Egypt, Sudan, and Ethiopia. The construction of the Grand Ethiopian Renaissance Dam (GERD) on the Blue Nile has been a point of contention between Ethiopia and Egypt, with Egypt fearing that reduced water flow from the dam will affect its already strained water supply. This dispute has heightened tensions over how to share the Nile's water resources (Elhadi et al., 2020).

Lake Chad Basin: Lake Chad, which is shared by Chad, Nigeria, Cameroon, and Niger, has been shrinking rapidly due to overuse, climate change, and diversion for irrigation. The shrinking lake has exacerbated tensions among the four countries, which depend on it for water, agriculture, and fishing (Olonisakin, 2019). There have been occasional clashes over water rights and control over access to the lake's remaining resources.

Zambezi River Basin: The Zambezi River is shared by eight countries, and competition for its waters has led to political disputes, especially in light of droughts and changing weather patterns. Zimbabwe, Zambia, and Mozambique all depend on the river for irrigation, hydropower, and drinking water, which leads to competition for water during periods of low rainfall (Chenje & Johnson, 2017).

3.10.1.6 Impacts of water conflicts

3.10.1.6.1 Cross-border tensions

Water-related disputes often contribute to diplomatic and military tensions between countries. The competition for shared resources like the Nile or Lake Chad can hinder regional cooperation on broader issues such as trade, security, and development.

Water scarcity and competition over shared water resources often exacerbate diplomatic and military tensions between neighboring countries.

These disputes can transform vital rivers, lakes, and aquifers into sources of conflict rather than cooperation. For instance, the Nile River, which flows through multiple countries including Ethiopia, Sudan, and Egypt, has been a focal point of regional friction for decades. Ethiopia's construction of the Grand Ethiopian Renaissance Dam (GERD) has heightened tensions with Egypt, which relies on the Nile for more than 90% of its freshwater needs. Egypt fears that the dam will significantly reduce the river's flow, threatening its agricultural sector and drinking water supplies. Similarly, the shrinking waters of Lake Chad have led to competition among Nigeria, Chad, Cameroon, and Niger. The lake, which was once one of Africa's largest, has diminished by over 90% since the 1960s due to a combination of climate change, overuse, and mismanagement. The resulting scarcity has intensified conflicts among communities dependent on its resources, contributing to regional instability.

These water-related disputes can spill over into other areas, impeding broader regional cooperation on crucial issues such as trade, security, and sustainable development. Rather than collaborating to develop joint water management strategies, countries may resort to aggressive policies, military posturing, or unilateral projects that exacerbate tensions. The lack of effective transboundary water management agreements undermines the potential for diplomatic resolution, making it difficult to build trust and cooperative frameworks in other areas of mutual interest. Thus, water scarcity threatens environmental sustainability and poses significant risks to geopolitical stability, particularly in regions already vulnerable to conflict.

3.10.1.6.2 Humanitarian impacts

The humanitarian consequences of water conflicts are profound, often leading to displacement, infrastructure destruction, and increased poverty. When access to clean water is restricted due to disputes, entire communities are forced to relocate, losing their homes, livelihoods, and access to basic services. In regions such as the Middle East and Sub-Saharan Africa, where water scarcity is already a pressing issue, conflicts over limited resources exacerbate social vulnerabilities. For example, the Yemen crisis demonstrates how water scarcity can deepen humanitarian suffering. Prolonged conflict has not only damaged water infrastructure but also limited access to safe drinking water, increasing the risk of waterborne diseases like cholera. This situation disproportionately affects vulnerable populations, including children and the elderly, exacerbating already dire living conditions.

Furthermore, the destruction of water infrastructure during conflicts leaves communities without the means to access clean water, which in turn disrupts agricultural activities, exacerbates food insecurity, and contributes to rising poverty levels. Without reliable water supplies, farming communities cannot sustain their crops, leading to reduced food production and higher prices. This chain reaction of water scarcity, agricultural decline, and economic hardship often triggers a cycle of poverty that can be difficult to break. As water becomes increasingly scarce, competition intensifies not only between nations but also within communities, leading to local conflicts that further strain social cohesion and stability. Thus, the humanitarian impacts of water conflicts extend far beyond immediate resource scarcity, affecting the social fabric, economic prospects, and long-term resilience of affected populations. Water conflicts can lead to the displacement of communities, destruction of infrastructure, and increased poverty as access to clean water becomes restricted during times of crisis (Gassert et al., 2020).

3.10.2 Severe agricultural limitations

Agriculture is a vital sector in Sub-Saharan Africa, with a large portion of the population depending on it for livelihoods. However, water scarcity has led to severe agricultural limitations, threatening food security and the region's economic stability.

3.10.2.1 Challenges to agriculture

Droughts and Erratic Rainfall: Many countries in Sub-Saharan Africa experience frequent droughts due to erratic rainfall patterns exacerbated by climate change. This greatly reduces the availability of water for rain-fed agriculture, which is the dominant agricultural practice in the region (Thornton et al., 2017). This vulnerability to drought has led to crop failures, food shortages, and reduced agricultural productivity.

Overreliance on Irrigation: In areas where irrigation is used, water shortages have reduced irrigated areas as water becomes more difficult to access. For example, in South Africa, where irrigation is crucial for farming, farmers need help with their choices as water resources dwindle, threatening crops like maize and wheat (Pereira et al., 2019).

Declining Groundwater Levels: In some parts of Sub-Saharan Africa, groundwater, which supports agriculture, is being extracted faster than replenished. Over-extraction of groundwater for irrigation in places like Botswana and Kenya has significantly reduced water availability, making it harder to sustain agricultural production in the long term (Tshikhudo et al., 2021).

3.10.2.2 Impacts on agriculture

Food Insecurity: Reduced agricultural production leads to food shortages and higher food prices, making it harder for people to afford basic staples. Many countries in Sub-Saharan Africa rely on agriculture for food, and water scarcity directly threatens the region's food security.

Economic Decline in Rural Areas: Agriculture is a key source of employment and income in rural Sub-Saharan Africa. As agricultural yields decrease due to water scarcity, millions of rural households are plunged deeper into poverty. For example, Ethiopia's water scarcity has reduced agricultural productivity, pushing many farmers into economic hardship (Bachewe et al., 2020).

3.11 Sub-Saharan Africa: Variable Climate, Poor Infrastructure, and High Dependence on Rainfall

Sub-Saharan Africa faces unique challenges in water management due to its variable climate, poor infrastructure, and high dependence on rainfall. These factors collectively contribute to the region's vulnerability to water scarcity and climate variability, significantly impacting agricultural productivity, economic development, and human well-being. Understanding the relationship between these issues is essential for developing sustainable solutions to address the water crisis in the region.

3.11.1 Variable climate and its impacts

One of the defining features of Sub-Saharan Africa is its variable climate, characterized by irregular rainfall patterns, frequent droughts, and unpredictable seasonal variations. Climate variability is particularly pronounced in the Sahel region, located just south of the Sahara Desert, where rainfall is low and erratic. The region's rainfall patterns are highly sensitive to phenomena such as El Niño and La Niña, which can exacerbate droughts or lead to excessive rainfall and flooding (Boko et al., 2007). This variability poses significant challenges for water resource management as traditional rainfall patterns, which local communities rely on for agriculture and drinking water, are becoming increasingly unreliable.

For instance, in Ethiopia, Somalia, and Chad, recurrent droughts have severely impacted agricultural production, livestock health, and water availability, leading to food insecurity and mass displacement (Kiros et al., 2017). These challenges are expected to intensify with ongoing climate change, further complicating efforts to address water scarcity. The climate's unpredictability has made it difficult for governments and communities to plan effectively for future water needs, as infrastructure and resource management systems are not always adaptable to such fluctuations.

3.11.2 Poor infrastructure and its limitations

Sub-Saharan Africa's water scarcity is further exacerbated by poor water storage, distribution, and treatment infrastructure. Water supply systems in many parts of the region are outdated, insufficient, or non-existent, particularly in rural areas. According to the African Development Bank (AfDB), poor infrastructure is one of the primary barriers to improving access to clean water, with only 60% of rural populations having reliable access to water sources (AfDB, 2016). In urban centers, infrastructure may exist but is often overwhelmed due to rapid population growth and inadequate maintenance.

In many countries, the need for reliable water distribution networks and storage facilities forces communities to depend on reliable sources, such as shallow wells or unprotected surface water bodies, which are often contaminated. This results in widespread waterborne diseases and poor sanitation conditions, leading to health crises that further strain limited resources. Furthermore, infrastructure investments need to be more robust due to financial constraints, political instability, and competing development priorities, leaving many communities vulnerable to water shortages.

3.11.3 High dependence on rainfall for agriculture

In Sub-Saharan Africa, agriculture is a major economic activity, which is heavily dependent on rain-fed agriculture. This sector employs most of the population, and its success is intimately tied to rainfall patterns. However, the high dependence on rainfall for crop production makes agriculture extremely vulnerable to climate change and variability. In Kenya, Mozambique, and Tanzania, droughts and irregular rainfall significantly affect crop yields, leading to food shortages and economic losses (FAO, 2020).

Reliance on rainfall also limits the development of irrigation systems, which are essential for ensuring reliable crop production in areas with inconsistent rainfall. While some countries, such as South Africa, have invested in large-scale irrigation infrastructure, many others need more resources and technological capacity to develop such systems. This contributes to low agricultural productivity, poor food security, and economic instability in a region where the population is rapidly growing and the demand for food is increasing.

Moreover, the absence of modern irrigation techniques means that farmers are forced to rely on traditional practices, which are less efficient and more susceptible to the changing climate. Soil degradation, deforestation, and inadequate agricultural policies further compound the situation, all of which contribute to reduced water retention and soil fertility, further limiting agricultural productivity (Nhamo & Ndlela, 2019).

3.11.4 Addressing water scarcity in Sub-Saharan Africa

A multifaceted approach is required to address the challenges posed by climate variability, poor infrastructure, and high dependence on rainfall. First, improving water infrastructure is essential to increasing access to clean water and improving storage capabilities. Investments in modern water treatment plants, rainwater harvesting, and water recycling systems can help ensure that water resources are better managed, even in the face of climate variability.

Additionally, promoting sustainable agricultural practices is critical for reducing the dependence on rain-fed agriculture. This includes implementing drip irrigation, improving soil health through conservation agriculture, and enhancing water management techniques at the farm level (Chiverge et al., 2017). Furthermore, regional cooperation on water management, through initiatives like the African Water Vision 2025 and Transboundary Water Management (TWM) frameworks, can help ensure equitable distribution of water resources across borders and promote shared solutions to common water challenges.

Lastly, enhancing climate resilience through adaptation strategies such as early warning systems, drought-resistant crops, and community-based water management can help mitigate the effects of water scarcity and climate change on vulnerable populations (Mwandiga, 2021).

Sub-Saharan Africa's vulnerability to water scarcity is driven by a combination of variable climate, poor infrastructure, and high dependence on rainfall for agricultural and domestic needs. Addressing these challenges will require significant investment in water infrastructure, enhanced agricultural practices, and international cooperation. By taking a holistic approach that includes improving water access, adopting sustainable farming techniques, and building climate resilience, Sub-Saharan Africa can begin to tackle its growing water crisis and move toward a more sustainable and equitable future.

3.12 Impact of Water Scarcity in Sub-Saharan Africa: Frequent Droughts, Health Issues from Poor Water Quality, and Economic Instability

Water scarcity in Sub-Saharan Africa profoundly impacts the region, affecting human health, economic stability, and overall social well-being. Due to a combination of climate variability, poor infrastructure, and high dependence on rain-fed agriculture, the region faces unique challenges that manifest in frequent droughts, health issues from poor water quality, and widespread economic instability. These challenges are exacerbated by growing population pressures, climate change, and political challenges, making the situation more complex and difficult to address. Below is a detailed discussion of how these issues impact Sub-Saharan Africa.

3.12.1 Frequent droughts and their effects

One of the most severe manifestations of water scarcity in Sub-Saharan Africa is the frequent occurrence of droughts. Many parts of the region, especially the Sahel and Horn of Africa, experience recurring droughts worsening due to climate change (Boko et al., 2007). These droughts reduce water availability for agriculture, livestock, and domestic use, leading to severe food and water insecurity. In countries like Ethiopia, Somalia, and Kenya, droughts have resulted in crop failures, livestock deaths, and a collapse in local food systems, which further exacerbate poverty and displacement (Kiros et al., 2017).

For example, Ethiopia's recurring droughts in the past decades, often linked to El Niño and La Niña events, have caused widespread crop failure and significantly affected the country's ability to provide for its population. These droughts, combined with water scarcity, also increase competition for limited water resources, often leading to conflicts between neighboring communities or countries (Kiros et al., 2017). As a result, the agricultural sector, which relies heavily on seasonal rainfall, is severely impacted, thus undermining food security and economic resilience.

3.12.2 Health issues from poor water quality

Access to safe drinking water remains a critical issue in Sub-Saharan Africa. Due to poor infrastructure, many communities rely on unsafe water sources, such as shallow wells, rivers, and unprotected springs, often contaminated by pollutants, bacteria, and viruses. This situation exacerbates waterborne diseases such as cholera, diarrhea, and dysentery, which are leading causes of morbidity and mortality in the region (WHO, 2014). The lack of clean water and poor sanitation leads to a cycle of poor health, further hampers development in many countries.

In countries like Niger and Nigeria, inadequate access to safe water has caused recurrent outbreaks of cholera and other waterborne diseases, resulting in high healthcare costs, social disruption, and loss of life (Bachmann et al., 2014). These health challenges also contribute to the burden on local health systems, which are often underfunded and understaffed, making it difficult for the population to receive timely and adequate care.

Moreover, the gendered impacts of water scarcity are noteworthy, as women and girls often bear the responsibility of water collection, which can lead to physical exhaustion and health risks, particularly when water sources are far from home. In many regions, women and children are forced to walk long distances to fetch water from unsafe sources, exposing them to dangerous health risks.

3.12.3 Economic instability due to water scarcity

Water scarcity significantly impacts the economic stability of Sub-Saharan African nations. Agriculture is a major economic driver in the region, and it is heavily dependent on rain-fed water systems. When droughts occur, crop yields drop, leading to food shortages, inflation, and rural poverty. In addition, water scarcity also affects the livestock sector, particularly in countries like Mauritania, Chad, and Ethiopia, where pastoralist communities depend on access to water for grazing and drinking. As a result, livestock deaths and losses in productivity have caused major economic setbacks in these regions (FAO, 2020).

In addition to agriculture, industry and energy production are also impacted by water scarcity. Many countries in Sub-Saharan Africa rely on hydropower for electricity, but with reduced water levels in reservoirs, power generation is affected, leading to power outages and increased energy costs. For example, Ethiopia's Grand Renaissance Dam, a major hydropower project, has faced significant water level challenges during dry periods, affecting its ability to generate power (Hickman et al., 2019). These disruptions in electricity production undermine economic activities, reduce industrial output, and slow down economic growth.

The economic instability caused by water scarcity also creates poverty cycles that are difficult to break. As agricultural production falls, food prices rise, and the cost of living increases, particularly for the urban poor who may not have direct access to rural resources. This economic insecurity also leads to migration, as people in drought-affected areas move to urban centers or neighboring countries in search of better opportunities, putting pressure on urban infrastructures and services (AfDB, 2016).

Water scarcity in Sub-Saharan Africa has far-reaching consequences, including frequent droughts, health issues from poor water quality, and economic instability. These challenges are deeply intertwined with the region's climate variability, lack of infrastructure, and high dependence on agriculture. Addressing these issues requires a multi-pronged approach, including investment in water infrastructure, development of climate-resilient agricultural practices, and regional cooperation on water resource management. Moreover, improving access to safe water and sanitation is essential for protecting public health and reducing the burden of waterborne diseases. Without addressing these core issues, Sub-Saharan Africa will continue to struggle with the negative effects of water scarcity, undermining efforts toward sustainable development and poverty alleviation.

3.13 South Asia-water Scarcity

Water scarcity is one of the most pressing challenges facing Asia, a continent that accounts for more than 60% of the world's population. The continent's vast geographical diversity and significant variations in climate, population density, industrialization, and water management practices contribute to the uneven distribution of water resources. Asia's water crisis is not just a regional issue; it is a global one that demands urgent attention due to its far-reaching environmental, economic, and social implications. While some countries in Asia face a relative abundance of freshwater resources, others are experiencing severe water stress, which threatens agriculture, industry, and the livelihoods of millions of people.

In South Asia, the problem of water scarcity is compounded by rapid population growth, high agricultural water demand, and deteriorating water quality. India, with its burgeoning population, is particularly vulnerable, as it experiences extreme regional disparities in water availability. Northern India, for example, relies heavily on the over-extraction of groundwater, which has led to significant declines in water tables and worsening water quality (Kummu et al., 2016). Neighboring Pakistan faces similar challenges, with agricultural activities heavily dependent on the Indus River, a transboundary water source, which is under strain from both upstream and downstream water demands (Ali et al., 2021). The competition for water, both within countries and across borders, increases the risk of conflicts over water resources, further complicating efforts to address water scarcity in the region.

Southeast Asia, though endowed with abundant water resources, faces its own set of challenges, including the impacts of climate change, infrastructural deficits, and dam construction on major river systems. The Mekong River, which supports the livelihoods of millions in countries like Vietnam, Cambodia, and Laos, is increasingly at risk due to upstream dam projects that disrupt the natural flow of water and threaten aquatic ecosystems (Mekong

River Commission, 2020). Meanwhile, urbanization and industrialization in countries like Thailand and Indonesia have significantly increased water demand, while many areas still rely on antiquated infrastructure, leading to inefficiencies and inadequate access to clean water for large segments of the population (Satterthwaite et al., 2018).

In Central Asia, water scarcity is exacerbated by the arid climate and the heavy reliance on transboundary rivers such as the Amu Darya and Syr Darya. The region's five countries—Kazakhstan, Kyrgyzstan, Tajikistan, Turkmenistan, and Uzbekistan—face significant challenges in managing these shared water resources. Political disputes and poor management have hindered effective cooperation, exacerbating the water crisis and threatening regional stability (Timmerman et al., 2017). These countries also face challenges in maintaining agricultural productivity, as their economies are highly dependent on irrigation systems that are unsustainable and inefficient.

The Middle East, particularly countries like Saudi Arabia, Iraq, and Iran, suffers from extreme water scarcity due to its low levels of renewable water resources, compounded by political and geopolitical conflicts. Water resources are often shared across borders in these countries, leading to competition and conflict over rivers like the Tigris and Euphrates (Allan, 2002). The reliance on non-renewable water sources, such as fossil groundwater and desalinated water, has further strained the region's ability to maintain a sustainable water supply (El-Beltagy et al., 2018). The impact of climate change, which leads to higher temperatures and lower precipitation, is expected to worsen these challenges, making water management an increasingly urgent issue.

China, the largest country in Asia by both population and land area, faces a complex set of challenges related to water scarcity, particularly in its northern regions. Despite being home to the Yangtze and Yellow Rivers, which provide water to millions, the country's water resources are unevenly distributed, with the northern regions suffering from chronic water shortages (Zhang et al., 2017). Rapid industrialization and urbanization have further increased water demand, while pollution and inefficient agricultural practices have diminished water quality. In response, China has invested heavily in large-scale water transfer projects, such as the South-to-North Water Diversion Project, which aims to redistribute water from the south to the arid north (Ma et al., 2020). However, these projects come with their own set of environmental and social concerns, including the displacement of communities and potential ecological damage.

In summary, water scarcity in Asia is a complex, multifaceted issue driven by climatic, geographical, political, and socioeconomic factors.

Ensuring equitable access to water across the continent requires a nuanced understanding of these dynamics and concerted local, national, and regional efforts to improve water management practices, increase efficiency, and address the root causes of water scarcity. Effective solutions must prioritize cooperation over conflict, sustainable water management strategies, and innovative approaches to addressing both supply and demand for water resources.

3.13.1 Asia: Variable climate, poor infrastructure, and high dependence on rainfall

Asia, a continent with diverse climates, faces significant water scarcity challenges, primarily due to variable climate conditions, poor infrastructure, and a high dependence on rainfall for freshwater resources. These factors create an uneven water distribution, leading to severe water stress in certain regions, while others experience water abundance but struggle with management issues.

3.13.1.1 Variable climate

Asia's climate varies widely from the monsoonal regions of South and Southeast Asia to the Middle East and Central Asia arid zones. Countries like India, Pakistan, and Bangladesh experience significant seasonal rainfall variability, with most precipitation occurring during the monsoon season. However, rainfall distribution is often irregular, leaving some regions with excessive water during the rainy season and severe shortages during dry periods (Kummu et al., 2016). In contrast, Central Asia and the Middle East are primarily arid or semi-arid, where rainfall is scarce, and water availability depends on transboundary rivers and non-renewable groundwater (Gleick, 2014). As climate change exacerbates these fluctuations, the variability in precipitation is expected to worsen, making water resources less predictable and more difficult to manage (Choudhury et al., 2017).

3.13.1.2 Poor infrastructure

The region also needs to improve its water infrastructure. In many countries, particularly in South Asia and Sub-Saharan Africa, existing water supply systems are outdated or insufficient. Poor infrastructure results in inefficiencies in water distribution, high levels of water loss, and inadequate access to clean water for large segments of the population. In rural areas, reliance on traditional water collection methods, such as wells and rainwater harvesting, can exacerbate water scarcity, especially during droughts (Satterthwaite et al., 2018). In urban areas, rapid population growth and inadequate investment in water infrastructure strains existing systems, leading to water shortages, contamination, and inequality in water access. For example, in India, inadequate water storage, treatment, and delivery systems have hindered the country's ability to address the growing water crisis (Kummu et al., 2016).

3.13.1.3 High dependence on rainfall

Water availability in many parts of Asia highly depends on seasonal rainfall, especially in countries with large agricultural sectors. Countries such as India, Pakistan, and China rely heavily on rain-fed irrigation systems to support their agriculture. However, rainfall-dependent agriculture is becoming increasingly unsustainable with increasing frequency and intensity of droughts (Ali et al., 2021). The reliance on rainfall for crop production and the depletion of groundwater resources due to over-extraction exacerbate food and water insecurity in these regions. In Southeast Asia, the Mekong River system is crucial for agriculture, but upstream dams and climate change are reducing water flows, further threatening food production and water availability (Mekong River Commission, 2020).

Asia's water scarcity challenges are multifaceted and influenced by climate variability, inadequate infrastructure, and overreliance on rainfall. A combination of strategies is needed to address these issues, including improving infrastructure, adopting more efficient irrigation techniques, and investing in climate-resilient water management systems. Collaborative efforts, particularly in regions with transboundary water sources, are essential to ensure sustainable water use for both current and future generations.

Water scarcity in Asia has far-reaching implications, particularly in terms of the reliance on desalination, conflicts over water resources, and severe limitations on agriculture. Each factor plays a significant role in shaping the region's economic, social, and political landscape.

3.13.1.4 Reliance on desalination

In regions where freshwater sources are scarce, countries have turned to desalination as a solution to meet their growing water demands. The Middle East and parts of Asia are among the world leaders in desalination technology, with countries like Saudi Arabia, the United Arab Emirates, and Singapore investing heavily in desalination plants. In Asia, countries such as Israel, Singapore, and parts of China have expanded desalination capacities to support both domestic and industrial water needs. For example, Singapore's highly advanced desalination and water recycling programs have been crucial in ensuring water supply despite limited natural freshwater resources (Lazarova & Bahri, 2017). However, desalination comes with challenges, including high energy consumption, environmental impacts from brine disposal, and high costs, which limit its feasibility in less economically advanced nations.

3.13.1.5 Conflict over water resources

Water scarcity in Asia has also exacerbated tensions between countries over shared water resources. The region is home to several major rivers that flow across national boundaries, such as the Mekong, the Ganges, and the Indus rivers. These rivers are crucial to the water supply of multiple countries, making them sources of geopolitical conflict. In the case of the Mekong River, for instance, upstream dam construction in China has significantly altered water flows, causing downstream countries like Cambodia, Laos, and Vietnam to express concerns about reduced water availability for agriculture and fishing (Mekong River Commission, 2020). Similarly, India and Pakistan have a long history of conflict over the Indus River, exacerbated by both water scarcity and political tensions. Water-sharing agreements have been made, but the ongoing political instability in the region continues to threaten the effective management of these critical water resources (Gleick, 2014).

3.13.1.6 Severe agricultural limitations

Agriculture, a key sector in many Asian economies, is highly dependent on water availability, and water scarcity has severe implications for food security and economic stability. In countries like India, Pakistan, and China, agriculture relies heavily on irrigation, with most of the irrigation water coming from groundwater, rivers, or reservoirs. However, crop yields suffer as water sources deplete due to over-extraction, groundwater levels drop, and rivers become unreliable during droughts. In India, for example, the Punjab region, often called the "Granary of India," faces severe water scarcity, leading to unsustainable farming practices and the over-extraction of groundwater for irrigation (Bierkens, 2015). In Central Asia, water diversion for cotton farming has led to the depletion of the Aral Sea, resulting in severe environmental and agricultural consequences (Pahl-Wostl et al., 2013). Climate change further complicates this situation by reducing the predictability of rainfall patterns, making farming even more vulnerable to water shortages.

The impact of water scarcity in Asia is multifaceted, with reliance on desalination, conflicts over shared water resources, and agricultural limitations all playing crucial roles in the region's struggle to manage water effectively. As the population grows and climate change accelerates, Asian countries must develop integrated water management strategies that promote cooperation, improve infrastructure, and invest in sustainable agricultural practices to address the growing water crisis.

3.13.2 Countries: Pakistan, India (with a focus on Rajasthan and Gujarat), and parts of Afghanistan

The South Asian region, which includes countries like Pakistan, India, Afghanistan, and others, is grappling with severe freshwater scarcity due to a combination of factors, including rapid population growth, poor water management, and climate change. This issue is particularly pronounced in South Asia's arid and semi-arid regions, including Rajasthan and Gujarat in India and parts of Afghanistan and Pakistan, where natural water resources are limited, and the demand for water is rising steadily.

3.13.2.1 Pakistan

Pakistan is one of the world's most water-scarce countries, with most of its water resources concentrated in the Indus River Basin. However, the country faces severe challenges related to water scarcity, including over-extraction of groundwater, inefficient irrigation systems, and pollution. A key issue in Pakistan is the over-reliance on the Indus River, which is shared with India. Both countries have a long history of tension and conflict over managing and allocating this vital water resource. As a result, water scarcity is not only a local environmental issue but also a geopolitical one (Gleick, 2014). Furthermore, Pakistan faces the dual challenge of decreasing water availability due to both climate change and rising demand for agricultural and industrial use (Khan & Ghaffar, 2015).

3.13.2.2 India: Rajasthan and Gujarat

India, particularly in the arid states of Rajasthan and Gujarat, faces significant water stress. Rajasthan, which is located in India's northwestern region, is one of the driest states, with most of its water supply coming from seasonal monsoon rains, which are highly unpredictable due to climate change. The state suffers from insufficient groundwater recharge, and in many rural areas, people rely on groundwater that has been severely over-exploited, leading to falling water tables (Shah, 2014). Gujarat, another state in western India, also faces similar issues with both groundwater depletion and high evaporation rates, exacerbating water scarcity. Despite these challenges Gujarat has seen some success with water management projects, such as the Sardar Patel Project, which has focused on improving irrigation systems and increasing the efficiency of water usage (Gupta & Singh, 2018).

India's Punjab region, often referred to as the country's "breadbasket, "faces severe water scarcity due to over-exploitation of groundwater. The region has a high population density and extensive agricultural activities, particularly the cultivation of water-intensive crops like rice and wheat. These crops and intensive irrigation practices have led to a significant depletion of groundwater resources. According to Bierkens (2015), Punjab is one of the most water-stressed regions in India, with groundwater levels declining rapidly due to overuse.

Moreover, the region's dependence on the monsoon for irrigation has made it vulnerable to fluctuations in rainfall patterns. The combined effect of over-extraction, climatic changes, and high demand for water from both agriculture and a growing population has made sustainable water management a pressing challenge. The situation in Punjab is a clear example of how the industrialization of agriculture can lead to water scarcity, even in regions that once had relatively abundant freshwater resources.

3.13.2.3 Afghanistan

Afghanistan is another South Asian country facing extreme water scarcity, primarily due to its semi-arid climate, poor infrastructure, and political instability. The country relies heavily on the Amu Darya and Helmand rivers, both of which are shared with neighboring countries, raising concerns about transboundary water disputes (Liu et al., 2015). Moreover, Afghanistan's water resources are unevenly distributed, with some regions experiencing chronic droughts while others face severe flooding. This inconsistency and poor water management systems have led to a water crisis that hampers agricultural productivity and contributes to food insecurity (Bakker, 2012).

3.13.3 Key factors contributing to freshwater scarcity

Several factors contribute to freshwater scarcity in the South Asian region. Climate variability, marked by erratic monsoon patterns and frequent droughts, is a significant challenge for countries like India, Pakistan, and Afghanistan. High evaporation rates, especially in arid regions like Rajasthan, further exacerbate the situation, as water losses are higher than the region's natural replenishment rates. Additionally, poor infrastructure, such as inefficient irrigation systems and inadequate water storage facilities, limits the ability of these countries to manage and distribute available water resources effectively. The high dependence on rainfall for agricultural irrigation, particularly in rural areas, makes the region especially vulnerable to fluctuations in precipitation patterns due to climate change.

Water scarcity in South Asia is a multifaceted issue with wide-ranging consequences for economic development, social stability, and food security. In countries like Pakistan, India (with a focus on Rajasthan and Gujarat), and Afghanistan, the impacts of water scarcity are intensified by a combination of natural and human-induced factors. Policymakers in these countries must address both the immediate and long-term challenges posed by water scarcity through enhanced water management strategies, regional cooperation on transboundary waters, and increased investments in infrastructure.

3.13.4 Causes: High population, agricultural demands, climate change, and weak water management infrastructure

Asia, home to over 60% of the global population, faces significant challenges related to freshwater scarcity due to many factors, including high population density, agricultural demands, climate change, and weak water management infrastructure.

3.13.4.1 High population

Asia's rapidly growing population puts immense pressure on the continent's water resources. Countries such as China, India, and Indonesia are home to some of the largest populations in the world, and water demand for drinking, sanitation, and industrial use has surged in line with this growth. The increasing number of people leads to a greater strain on existing freshwater resources, further exacerbating scarcity issues. Population growth in urban areas has also resulted in high competition for water between domestic, industrial, and agricultural sectors, which limits the available water supply for sustainable use (Shah, 2014).

3.13.4.2 Agricultural demands

Agriculture is the largest consumer of water in Asia, accounting for roughly 80% of the region's freshwater use (Mekonnen & Hoekstra, 2016). The growing demand for food, driven by both population increases and changing dietary preferences, has significantly increased the need for irrigation. In countries like India and Pakistan, vast areas are irrigated using water from rivers and groundwater, placing a heavy burden on freshwater resources. Over-extraction of water for irrigation, coupled with inefficient irrigation practices, has led to the depletion of aquifers and reduced surface water availability, contributing to overall water scarcity (Shah, 2014). Climate change, which impacts the timing and distribution of rainfall, further complicates agricultural water use, making agriculture in some regions increasingly unsustainable.

3.13.4.3 Climate change

Climate change exacerbates water scarcity in Asia, particularly in South and Central Asia, where many countries are experiencing more frequent and severe droughts. Changes in precipitation patterns, such as reduced rainfall or erratic monsoons, have intensified the variability of water availability in countries like India, Pakistan, and Afghanistan. In many parts of Asia, the reduced snowmelt from mountain glaciers due to rising temperatures threatens river flows, further compounding water stress (Bakker, 2012). Sea-level rise, particularly in low-lying areas like Bangladesh and parts of Southeast Asia, also leads to the salinization of freshwater supplies, reducing the availability of potable water for drinking and irrigation (Mekonnen & Hoekstra, 2016).

3.13.4.4 Weak water management infrastructure

Despite the significant challenges posed by water scarcity, weak water management infrastructure is one of the key underlying causes of water stress in many Asian countries. Inadequate infrastructure for water storage, treatment, and distribution means that much of the available water is wasted or polluted. For example, in India, the large-scale inefficiency in irrigation systems, as well as water losses in municipal supply systems, results in large amounts of water being lost before it reaches its intended use (Bakker, 2012). Additionally, many countries in the region lack comprehensive water management strategies, and political and administrative inefficiencies further hinder efforts to address water scarcity at the national and regional levels.

Water scarcity in Asia is a multifaceted issue driven by high population growth, agricultural demands, climate change, and weak water management infrastructure. As Asia continues to grow, addressing these challenges will require more efficient use of water resources, investments in sustainable infrastructure, and regional cooperation to ensure equitable and sustainable water access for all.

3.13.5 Impact: Groundwater depletion, pollution, impacts on food security and livelihoods, regional water conflicts

In Asia, the impact of water scarcity is profound and multi-dimensional, influencing key sectors such as groundwater, agriculture, food security, livelihoods, and regional stability. The scarcity is increasingly linked to groundwater depletion, pollution, and ongoing regional conflicts over water resources.

3.13.5.1 Groundwater depletion

Groundwater is a vital resource for drinking water, agriculture, and industry in many parts of Asia. However, in countries like India, China, and Pakistan, excessive extraction for irrigation and urban consumption has led to significant depletion of groundwater aquifers. For instance, in the northwestern Indian states of Punjab and Rajasthan, groundwater levels have declined sharply due to intensive farming and unsustainable water practices (Shah, 2014). This depletion threatens the availability of water for future generations and leads to land subsidence and increased pumping costs, further exacerbating the water crisis. In many areas, over-extraction has made aquifers non-replenishable within human timescales (Mekonnen & Hoekstra, 2016).

3.13.5.2 Pollution of water sources

Water pollution is another critical issue affecting surface water and groundwater in Asia. Rapid industrialization, unregulated urbanization, and agricultural runoff have led to widespread contamination of freshwater sources. In countries like China and India, pollutants such as heavy metals, untreated sewage, and pesticides have rendered many rivers, lakes, and groundwater sources unsafe for consumption and agricultural use (Bakker, 2012). In particular, the Ganges River in India and the Yellow River in China are heavily polluted, affecting human health and undermining ecosystems that depend on clean water.

3.13.5.3 Impacts on food security and livelihoods

The water crisis in Asia directly impacts food security, as much of the continent's agriculture depends on irrigation from rivers and groundwater. Water scarcity and poor water management practices lead to crop failures and reduced agricultural productivity. Water shortages significantly threaten food security in regions such as the Indo-Gangetic Plain, where water-intensive crops like rice and wheat are grown. As farmers face dwindling water resources, they often resort to deepening wells or switching to less water-demanding crops, both of which can reduce income and exacerbate poverty (Shah, 2014). The agricultural sector in many Asian countries is also highly vulnerable to the impacts of climate change, such as altered monsoon patterns, which compound food insecurity (Mekonnen & Hoekstra, 2016).

3.13.6 Regional water conflicts

Regional water conflicts have become more frequent and intense, particularly in areas where multiple countries share transboundary river basins. One of the most well-known examples is the dispute over water sharing from the Mekong River, which affects China, Laos, Cambodia, Vietnam, and Thailand. China's construction of dams along the Mekong has raised concerns among downstream countries about reduced water flow and its implications for agriculture and livelihoods. Similarly, tensions between India and Pakistan over the Indus River, which is crucial for both countries' water supply, have been a longstanding source of conflict. Disputes over water resources have also escalated in the Middle East, especially between Turkey, Syria, and Iraq over the Euphrates and Tigris Rivers, where political tensions and military actions compound water scarcity (Bakker, 2012).

The impacts of water scarcity in Asia are severe and far-reaching, with consequences for groundwater depletion, pollution, food security, livelihoods, and regional peace. Addressing these challenges will require coordinated efforts to improve water management, reduce pollution, invest in sustainable irrigation practices, and enhance international cooperation on transboundary water resources.

3.13.6.1 California, USA

California is a well-known example of a region facing extreme water scarcity, particularly in the context of its agricultural industry and urban water demands. The state is home to one of the largest agricultural sectors in the United States, but its water resources are often insufficient to meet the needs of agriculture, industry, and its urban population. The competition for water among these sectors has led to conflicts over water allocation, especially during periods of drought (Postel, 2017).

The state's reliance on the Sierra Nevada snowpack for water storage has also created vulnerabilities. The snowpack is the primary source of freshwater during dry months, but it has been shrinking due to higher temperatures and reduced snowfall caused by climate change (Mount & Twiss, 2005). As a result, California faces the dual challenge of managing water scarcity while addressing the growing impacts of climate change on water availability.

California is a prime example of a region facing significant water scarcity. Various factors contribute to the state's ongoing water challenges, including high agricultural demands, climate variability, population growth, and limitations in water infrastructure.

3.13.6.1.1 Water scarcity factors in California

Agricultural Demands: California is a major agricultural producer, contributing a substantial portion of the nation's food supply. However, agriculture accounts for about 80% of the state's water use, largely for irrigation. Crops like almonds, rice, and grapes are water-intensive, placing a heavy demand on available water resources (Harter et al., 2018). The state's reliance on irrigation, especially in the Central Valley, exacerbates its vulnerability to water shortages.

Climate Variability and Droughts: California has a Mediterranean climate characterized by hot, dry summers and mild, wet winters. However, the state is increasingly experiencing more frequent and severe droughts, driven in part by climate change. These droughts deplete both surface water and groundwater supplies. The state relies on snowmelt from the Sierra Nevada mountains for a significant portion of its annual water supply, and with rising temperatures, the snowpack has been shrinking, leading to reduced water availability (Hughes et al., 2016).

Population Growth and Urban Demand: California is the most populous state in the U.S., and its urban water demands continue to rise. While major cities like Los Angeles and San Francisco have made significant strides in water conservation, the overall demand for water in these urban areas remains high, further straining the state's water systems (Postel, 2017).

Water Infrastructure and Management: Despite efforts to improve water efficiency, the state's infrastructure is aging and often needs help to handle the pressures of growing demands and changing conditions. The California State Water Project (SWP) and the Central Valley Project (CVP) were designed to move water from the north to the south. Still, these systems face challenges related to maintenance, outdated technology, and environmental concerns (Gleick, 2014).

3.13.6.1.2 Impact and adaptation strategies

To mitigate the impacts of water scarcity, California has implemented several strategies:

Water Conservation: Efforts to reduce water usage in agricultural and urban sectors, including rationing during droughts and incentives for efficient water use, have been crucial.

Desalination Projects: As freshwater sources become increasingly unreliable, desalination projects, such as the Carlsbad desalination plant, have been developed to provide an alternative source of drinking water (Gleick, 2014).

Water Recycling: California has advanced in wastewater recycling, particularly in urban areas like Los Angeles and Orange County, which treat and reuse water for non-potable uses, including irrigation and industrial processes (Lazarova & Bahri, 2017).

A combination of climate variability, agricultural water demand, population pressures, and infrastructure challenges drives California's water scarcity. While the state has made significant strides in implementing conservation and alternative water sources, its future remains uncertain, requiring continued innovation and collaborative management efforts.

3.14 Australia (Murray-Darling Basin)

Australia's Murray-Darling Basin is a region where extreme water scarcity has become a persistent problem due to a combination of factors, including over-allocation of water resources, climate change, and drought. The Murray-Darling River system, which spans several states in southeastern Australia, has been over-exploited for irrigation, leading to the depletion of water reserves and declining water quality (King et al., 2009). Climate change has exacerbated the situation, with prolonged droughts and reduced rainfall further reducing the availability of water.

The Australian government has been implementing reforms to improve water management in the basin, but the region still faces significant challenges in balancing agricultural, ecological, and urban water demands. The example of the Murray-Darling Basin highlights how water management practices, especially in water-scarce regions, can have long-term ecological and social impacts if not handled sustainably.

3.14.1 Water scarcity in Australia: The Murray-Darling Basin and desert areas

Australia is a country characterized by vast arid regions, and it faces significant water scarcity issues, particularly in areas like the Murray-Darling Basin and other desert regions. These areas are critical to the country's agricultural economy, yet they face challenges related to climate variability, overextraction of water, and poor water management.

3.14.1.1 Murray-Darling Basin

The Murray-Darling Basin, which spans four states and covers about one-seventh of Australia, is the most important water catchment area in the country. It is responsible for producing a large proportion of Australia's agricultural output, including rice, cotton, and fruit, which are heavily reliant on irrigation. However, the Basin has faced long-standing water scarcity issues driven by several factors:

Climate Variability: The Basin experiences highly variable rainfall patterns, with some areas receiving very little rain for extended periods. This has resulted in regular droughts, which have severely impacted water availability.

According to the Australian Bureau of Meteorology, the Basin has seen significant reductions in streamflow due to changing precipitation patterns, especially in recent decades (CSIRO, 2017).

Over-extraction: The water extracted from the Basin for irrigation and other uses exceeds the amount of natural replenishment, leading to environmental degradation. The overuse of water for irrigation has led to a decline in water quality, the salinization of soils, and reduced biodiversity (Davies & Field, 2015).

Management Challenges: While the Murray-Darling Basin Plan was introduced to address water scarcity and ensure equitable water distribution, challenges remain in balancing the needs of agriculture, urban populations, and the environment. Additionally, inconsistent enforcement and allocation practices have created tensions between states and water users (Tilmant & Bos, 2017).

3.14.2 Desert areas: The simpson and gibson deserts

In addition to the Murray-Darling Basin, Australia's desert regions also face severe water scarcity. The Simpson and Gibson Deserts, which cover much of the country's central and western areas, receive little to no rainfall and are inhospitable for agricultural development without external water sources.

Climate Extremes: These desert regions are characterized by extreme temperatures, high evaporation rates, and minimal precipitation. As a result, natural water sources like rivers and lakes are scarce and nonexistent, creating reliance on underground aquifers, which are depleting at unsustainable rates. For example, the Great Artesian Basin, which supplies much of the water for the desert areas, is being over-exploited, leading to concerns about the future availability of this water resource (Hancock et al., 2014).

Water Resource Management: In these areas, the infrastructure required to supply water is costly and resource-intensive. Desalination plants, such as those in coastal regions like Perth, help mitigate some water shortages, but they come with environmental and economic costs. These solutions are not always applicable to the desert interior, where access to groundwater is often the only viable option (Postel, 2017).

Impact on Livelihoods: Water scarcity in desert regions significantly affects indigenous populations and pastoral communities, whose livelihoods depend on sustainable water access for both personal consumption and livestock management. Climate change is exacerbating these challenges by altering rainfall patterns, increasing evaporation, and leading to more extreme droughts (van der Merwe et al., 2021).

Australia's water scarcity issues, exemplified by the challenges in the Murray-Darling Basin and desert regions, are complex and multifaceted. They are driven by climate variability, over-extraction, poor water management, and the impacts of climate change. While the government has made strides in addressing these issues through various policies and management plans, continued collaboration and innovation are needed to ensure water security for both people and ecosystems in the future.

3.15 Comparative Overview of Water Scarcity in Latin America (South America)

Water scarcity in Latin America is a multifaceted issue, influenced by a range of factors such as geography, climate variability, population growth, economic development, and governance. While South America as a whole is home to some of the largest freshwater resources globally, such as the Amazon River, many countries face significant challenges in water access and quality. The causes and impacts of water scarcity vary widely across the continent, with countries like Brazil, Chile, and Bolivia experiencing different levels of stress on their water resources.

3.15.1 Brazil

Brazil is often seen as a country with abundant water resources due to the vast Amazon River Basin, the largest freshwater system in the world. However, regional disparities exist. The northeastern part of Brazil, particularly the semi-arid regions of the Sertão, faces severe water scarcity due to low rainfall and high evaporation rates. In contrast, the southern and central regions are more water-rich.

3.15.1.1 Factors driving water scarcity

Water scarcity in many regions, particularly in Sub-Saharan Africa and parts of Latin America, is driven by climate variability, over-extraction, and environmental degradation. These factors are interlinked and have compounding effects that exacerbate the region's water crisis. Below, we explore each of these factors in detail:

3.15.1.2 Climate variability

Climate variability, especially in the form of recurring droughts, is a major driver of water scarcity in many parts of the world. In regions like north-eastern Africa and parts of South America, the frequency and intensity of droughts have increased due to climate change. The El Niño phenomenon, which causes significant changes in weather patterns, is a key contributor to these droughts, particularly affecting rainfall patterns in countries like Ethiopia, Kenya, and the Horn of Africa. During El Niño years, these regions often experience drier-than-normal conditions, leading to reduced water availability for both consumption and agricultural needs (IPCC, 2022). In Latin America, particularly in areas bordering the Amazon rainforest, the fluctuations in seasonal rainfall due to climate variability have significant impacts on water resources, as lower rainfall directly reduces river flow and water storage in reservoirs.

3.15.1.3 Over-extraction

Over-extraction of water, particularly for agricultural purposes, is another critical factor contributing to water scarcity. In many regions, especially in Sub-Saharan Africa and parts of Latin America, water is heavily used for irrigation to support the agricultural sector, which is the backbone of local economies. In countries like South Africa, the extraction of water for irrigation in arid regions has placed immense pressure on available water resources. Over-extraction depletes aquifers and reduces surface water levels, which leads to a diminished capacity to provide water for drinking, sanitation, and industrial uses. In countries such as Argentina and Brazil, the overuse of water resources for agricultural irrigation in the southern parts has resulted in water shortages during dry seasons. Furthermore, inefficient irrigation techniques and water-intensive crops exacerbate the problem, as large amounts of water are used unnecessarily, further depleting already limited resources (FAO, 2020).

3.15.1.4 Deforestation

Deforestation, particularly in the Amazon rainforest, significantly alters local and regional water cycles, contributing to water scarcity. The Amazon, known as the "lungs of the planet," is critical in regulating global and regional weather patterns. Trees and plants absorb and release water through transpiration, which helps maintain the water cycle by contributing to cloud formation and rainfall. However, large-scale deforestation for agriculture, logging, and infrastructure development has disrupted these processes. In regions of the Amazon where deforestation is most rampant, such as in Brazil, the loss of forests reduces rainfall. It alters river flow, exacerbating water scarcity in nearby communities and affecting agricultural productivity (Nobre, 2016). As the forests disappear, the natural regulation of the water cycle diminishes, leading to drier conditions in areas that were once lush and capable of sustaining abundant freshwater resources.

The interplay of climate variability, over-extraction, and deforestation significantly contributes to many countries' water scarcity challenges. The ongoing impacts of climate change, paired with unsustainable water use practices and environmental degradation, highlight the urgency of implementing integrated water resource management strategies that consider the interconnectedness of these factors. Long-term solutions will require global cooperation, sustainable agricultural practices, forest conservation efforts, and policies to mitigate climate change and alleviate the pressures on water resources.

3.15.1.5 Impact of water scarcity and contamination in Brazil

Water scarcity and contamination are increasingly becoming significant challenges in Brazil, especially in the northeast, where climate variability and industrial activities have compounded the existing water stress. These challenges have wide-ranging impacts on agriculture, energy production, and migration patterns, making it essential to address both the supply and quality of water to ensure sustainable development.

3.15.2 Crop failure and agricultural impact

In northeast Brazil, water scarcity is a critical factor contributing to crop failure and reduced agricultural productivity. The region, which is predominantly arid or semi-arid, is highly susceptible to droughts, particularly during the El Niño phenomenon, which disrupts rainfall patterns. Prolonged droughts result in soil moisture depletion, affecting the cultivation of staple crops such as beans, corn, and rice. In turn, the failure of crops not only leads to food insecurity for local communities but also impacts the broader economy, as agriculture is a key sector in the region. According to Marengo et al. (2017), water shortages in the northeast have also reduced irrigation water availability, making it more difficult for farmers to sustain their livelihoods. As crop yields decrease, increasing reliance on imported food increases, raising the cost of living and exacerbating poverty.

3.15.2.1 Reduced hydropower generation

Water scarcity in Brazil has also significantly impacted the country's energy sector, particularly in hydropower generation. Brazil relies heavily on hydroelectric power for electricity, with around 70% of its energy coming from hydropower sources. However, prolonged droughts and reduced river flow in regions such as the northeast and the central parts of the country have reduced water availability for hydropower plants (Marengo et al., 2017). As water levels in reservoirs drop, power generation capacity is constrained, which in turn leads to energy shortages, higher electricity prices, and the potential for power rationing. The energy crisis not only affects households but also industries, which depend on a stable energy supply for their operations. This, in turn, can lead to economic disruptions, affecting Brazil's overall growth and development.

3.15.2.2 Migration pressures

Water scarcity has also contributed to migration pressures in Brazil, particularly in the northeast. As water resources become increasingly scarce and agriculture fails to sustain local populations, many people are forced to migrate for better living conditions. This migration often increases strain on urban centers, particularly in the south and southeast of Brazil, where migrants seek employment and access to water and food resources. The influx of migrants to already overcrowded cities exacerbates social and economic tensions, contributing to poverty, unemployment, and inadequate access to basic services. Moreover, migration's social and environmental impacts can increase tensions between local communities and migrants, creating additional challenges for policymakers and social services (Marengo et al., 2017).

3.15.2.3 Contamination of freshwater sources

In addition to the challenges posed by water scarcity, Brazil is also grappling with the contamination of its freshwater sources. Industrial activities, mining, and untreated sewage are significant contributors to the pollution of water sources, particularly in the Amazon and other key river basins. The contamination of rivers, lakes, and groundwater has serious implications for public health and the environment. Industrial waste, heavy metals from mining operations, and agricultural runoff introduce harmful substances into the water supply, making it unsafe for consumption and irrigation The lack of proper waste treatment infrastructure in many parts of the country further exacerbates this issue. As a result, communities that rely on contaminated water sources face a higher risk of waterborne diseases, such as cholera and dysentery, which in turn increase healthcare costs and strain the country's public health system. Additionally, polluted water affects biodiversity and aquatic ecosystems, with long-term environmental consequences for the region's flora and fauna (UNESCO, 2019).

Water scarcity and contamination are two of Brazil's most pressing environmental challenges. These issues impact agriculture, energy production, and migration and pose significant risks to public health and environmental sustainability. Addressing water scarcity and pollution will require comprehensive solutions, including improved water management, investment in sewage and wastewater treatment infrastructure, and policies aimed at protecting and restoring ecosystems. Additionally, addressing the root causes of water scarcity, such as climate change and deforestation, will be critical to ensuring the long-term sustainability of Brazil's water resources.

3.16 Chile

Chile is one of the driest countries in South America, especially in the northern regions. The Atacama Desert is one of the driest places on Earth, receiving very little rainfall. Both natural factors and human activities drive Chile's water crisis. Despite the abundance of water in the south, the north faces severe water shortages due to over-reliance on surface water, climate change, and population growth.

3.16.1 Factors driving water scarcity in Chile

Chile, a country known for its diverse geography, ranging from the arid Atacama Desert in the north to the snowy Andes Mountains, faces significant challenges in managing its freshwater resources. Several factors contribute to Chile's water scarcity: climate change, poor water management practices, and the water-intensive mining industry. These challenges significantly affect agriculture, urban water supply, and economic stability.

3.16.1.1 Climate change and reduced snowpack in the andes

One of the most significant drivers of water scarcity in Chile is climate change, particularly the reduction of snowpack in the Andes Mountains, a critical source of freshwater for the country. The snowpack is a natural reservoir, slowly releasing water during the warmer months, sustaining river flows, and replenishing aquifers. However, rising temperatures and shifting precipitation patterns have led to a decline in snowfall, resulting in diminished snowpack levels. According to studies by Masiokas et al. (2020), the

Andes have experienced a significant reduction in snow accumulation over the past few decades, with some areas seeing a decrease of up to 40%. This reduction directly affects river flows, particularly in central Chile, which relies heavily on meltwater for agriculture, drinking water, and hydropower generation. The decrease in snowpack is not only affecting water availability but also increasing the region's vulnerability to droughts, which have become more frequent and severe in recent years (Garreaud et al., 2017).

3.16.1.2 Poor water management practices

Water management in Chile has long been a contentious issue, particularly regarding the allocation and regulation of water rights. The country operates under a unique water code established in 1981, which allows for the privatization of water rights. This market-based approach has led to the concentration of water rights in the hands of large agricultural producers and mining companies, often at the expense of small farmers and local communities. Poor management practices, especially in the agricultural sector, exacerbate water scarcity, as inefficient irrigation techniques lead to significant water losses. Agriculture accounts for nearly 70% of the country's water use, with much of it used for water-intensive crops such as avocados and grapes, which are exported to international markets (Donoso, 2018). The over-extraction of water for agricultural purposes has led to the depletion of aquifers and reduced river flows, particularly in regions such as the Central Valley, which is the agricultural heartland of the country. The lack of integrated water management policies that consider the needs of all stakeholders, including environmental flows, has further exacerbated the problem (Vergara, 2019).

3.16.1.3 The impact of mining on water resources

Chile is the world's largest producer of copper, and the mining industry plays a crucial role in its economy. However, mining is also one of the most water-intensive industries, especially in the arid northern regions where water is already scarce. Copper extraction processes, such as ore processing and dust suppression, require vast amounts of water, which are often sourced from local aquifers and rivers. The over-extraction of water by mining companies has led to the depletion of groundwater reserves and the drying up of surface water bodies, which in turn affects local communities and ecosystems (Oyarzún & Oyarzún, 2011). In addition to water consumption, mining activities also pose a risk of water contamination due to the release of heavy metals and chemicals used in the extraction process. Studies have shown that the high levels of water extraction by mining operations, especially in the Atacama Desert, have led to significant ecological impacts, including the reduction of wetlands and the loss of biodiversity (Larraín & Schaeffer, 2020).

The factors driving water scarcity in Chile are complex and interconnected, with climate change, poor water management, and the mining industry's demands playing significant roles. Addressing these challenges will require a comprehensive and integrated approach to water management that balances the needs of different sectors while ensuring the sustainability of water resources. Policymakers must consider revising the country's water code to promote more equitable water distribution and enhance water conservation efforts, particularly in agriculture and mining. Additionally, investing in technologies such as efficient irrigation systems and water recycling in mining operations could help alleviate some of the pressures on water resources. As climate change continues to impact water availability, Chile will need to adopt adaptive measures to protect its freshwater resources and ensure water security for its population and industries.

3.16.1.4 Impact of water scarcity in Chile

Water scarcity in Chile has profound and far-reaching impacts on urban and rural communities, influencing socio-economic dynamics and environmental sustainability. The challenges posed by limited water availability have intensified in recent years due to climate change, poor water management, and the growing demand from various sectors, leading to conflicts over water access and increasing water supply costs (Hernandez et al., 2020).

3.16.1.5 Conflicts and rising costs in urban and rural areas

In urban areas, water scarcity has led to rising costs for households and businesses, as utilities struggle to maintain adequate supply amidst dwindling resources. In Santiago, the capital city, residents have experienced periodic water rationing, especially during extended drought periods, which has driven up the cost of water. These rising costs disproportionately affect low-income households, exacerbating social inequalities. Rural areas, particularly in central and northern Chile, are even more vulnerable, where water is crucial for agriculture and livelihoods. The competition for limited water resources often leads to conflicts between different user groups, such as farmers, urban developers, and industrial users (Hernandez et al., 2020).

The conflict over water rights and allocation has been particularly intense in regions where agriculture and mining dominate. For instance, farmers in the Central Valley have reported declining water levels in wells and rivers, which has reduced their crop yields and income. This situation has created tensions between agricultural communities and mining companies, major water consumers in the country's northern parts (Donoso, 2018).

The unequal distribution of water rights and inadequate regulation have further fueled these conflicts, as powerful industries often have greater access to water resources than local communities.

3.16.1.6 Impact on agriculture and food security

Chile's agricultural sector is one of the hardest hit by water scarcity, especially in the central and northern regions where water shortages are most severe. These areas are vital for the country's fruit and vegetable production, which is exported to global markets. The limited water availability has forced many farmers to reduce their crop cultivation, affecting both the domestic food supply and export revenues. For example, the production of water-intensive crops like avocados, grapes, and berries has been significantly affected, leading to reduced exports and higher prices in international markets (Garreaud et al., 2017).

The impact on agriculture also raises concerns about food security, particularly in rural areas where communities rely on local food production. Water scarcity has made it difficult for small-scale farmers to maintain their operations, reducing agricultural output and threatening the livelihoods of those dependent on farming. The decline in agricultural productivity due to water shortages affects local food availability and contributes to inflation in food prices, further exacerbating the socio-economic challenges faced by vulnerable populations (Vergara, 2019).

The impacts of water scarcity in Chile are multifaceted, affecting urban and rural communities, agriculture, and overall socio-economic stability. Addressing these challenges requires comprehensive water management reforms prioritizing equitable access, efficient use, and sustainable conservation of water resources. Without such measures, Chile's water scarcity issues will likely worsen, posing a significant threat to its economic development and social cohesion.

3.17 Bolivia

Bolivia, a landlocked country, faces significant water challenges, particularly in the high-altitude areas of the Andes and the eastern lowlands. Water scarcity in Bolivia is influenced by climate change, glacial retreat, and population pressures in urban centers like La Paz.

3.17.1 Factors driving water scarcity in Bolivia

Environmental, climatic, and human factors influence water scarcity in Bolivia. The country's unique geography, which includes the Andes

Mountains, Amazon rainforest, and vast lowlands, makes its water resources highly variable and sensitive to climate change. The major factors contributing to water scarcity in Bolivia include glacial retreat, over-extraction of water resources, and increasing climate variability.

3.17.1.1 Glacial retreat

One of Bolivia's most significant factors driving water scarcity is the Andean glaciers' rapid retreat. These glaciers are a crucial source of freshwater for millions of Bolivians, especially during the dry season when other water sources become scarce. Glaciers such as the Chacaltaya and the Tuni-Condoriri, which feed into rivers supplying major cities like La Paz and El Alto, have been shrinking at an alarming rate due to rising temperatures (Bradley et al., 2021). This glacial meltwater is essential for sustaining agriculture, hydropower, and urban water supply. However, as these glaciers continue to retreat, the long-term availability of water is threatened, posing risks to both food security and the livelihoods of communities that depend on these resources.

The loss of glacier mass in the Andes is a local issue and part of a broader pattern of glacial retreat observed worldwide due to global warming. The decline in glacier volume reduces freshwater's natural storage, exacerbating water scarcity during prolonged dry spells. The impacts are particularly severe in the high-altitude regions of Bolivia, where communities rely heavily on meltwater for their daily needs (Rabatel et al., 2013).

3.17.1.2 Over-extraction of water resources

Another critical driver of water scarcity in Bolivia is the over-extraction of water for agricultural purposes. Agriculture is a major part of the Bolivian economy, accounting for significant employment and GDP contributions. However, this sector is also the largest consumer of water, using around 80% of the country's water resources. Intensive irrigation practices, particularly in the lowland regions and valleys, have depleted rivers, lakes, and underground aquifers. For example, Lake Poopó, Bolivia's second-largest lake, has dramatically shrunk due to a combination of water diversion for agriculture and prolonged droughts, causing significant ecological and economic damage (Orellana-Alvear et al., 2017).

The unsustainable groundwater extraction for agriculture has also led to falling water tables, threatening drinking water availability for rural communities. This overuse of water resources and poor management practices have led to water shortages and increased competition between agricultural users and other sectors, exacerbating conflicts over water rights (Romero-Mujalli et al., 2020).

3.17.1.3 Climate variability

Bolivia's water scarcity is further compounded by increasing climate variability, irregular rainfall patterns, and extreme weather events. The country experiences alternating cycles of droughts and floods, which have become more intense and unpredictable in recent years due to climate change. During the dry season, many parts of Bolivia suffer from severe water shortages, impacting urban and rural populations. In contrast, during the rainy season, heavy downpours can lead to flooding, damaging infrastructure, and disrupting water supply systems (Seiler et al., 2013).

The effects of climate variability are particularly challenging for Bolivia's agricultural sector, which is highly dependent on rainfall. Irregular precipitation can result in crop failures, reduced yields, and increased food insecurity, particularly in vulnerable communities. The combination of prolonged droughts and sudden floods not only disrupts agricultural production but also threatens the country's water infrastructure, leading to increased costs for maintenance and repairs (Vergara et al., 2016).

Bolivia's water scarcity is driven by a complex interplay of glacial retreat, over-extraction of water for agriculture, and climate variability. Addressing these challenges requires comprehensive water management strategies incorporating sustainable agricultural practices, climate adaptation measures, and policies to protect critical water sources like the Andean glaciers. Without these efforts, Bolivia may face increasing water insecurity, seriously affecting its economic stability and social well-being.

3.17.1.4 Impact of water scarcity in Bolivia

Water scarcity in Bolivia has profound implications for both rural and urban populations, affecting agriculture, livelihoods, and access to potable water. These impacts are exacerbated by the country's dependency on seasonal water sources, which are increasingly threatened by climate change and poor resource management.

3.17.1.5 Impact on rural communities

Bolivia's rural communities rely heavily on agriculture, which is the primary source of income and sustenance. However, water shortages directly affect crop yields, leading to reduced agricultural productivity and, consequently, affecting the livelihoods of smallholder farmers. This is particularly critical in regions such as the Altiplano and the inter-Andean valleys, where agriculture is rain-fed and thus highly sensitive to changes in water availability (Vera, 2019). The depletion of water sources like rivers and aquifers, coupled with unpredictable rainfall patterns, exacerbates the vulnerability of these communities, leading to food insecurity and economic instability.

Water scarcity in rural areas also often results in increased competition for limited water resources, leading to conflicts between user groups. This situation is particularly challenging in areas where water is shared among agricultural, domestic, and industrial sectors. The lack of efficient water management infrastructure further compounds these challenges, making it difficult for rural communities to adapt to changing environmental conditions.

3.17.1.6 Impact on urban water systems

Urban centers in Bolivia, especially the capital city La Paz, are also facing significant challenges related to water scarcity. The city relies on a combination of glacial meltwater and seasonal rainfall to supply its water needs. However, the retreat of Andean glaciers, due to rising temperatures, has reduced the availability of these crucial water sources. This has led to periodic water shortages, particularly during the dry season, affecting the availability of potable water for urban residents (Hernandez et al., 2020).

The situation in La Paz was notably severe during the 2016-2017 water crisis, where prolonged drought led to water rationing and the declaration of a national emergency. The water shortage forced residents to queue for hours to access water from tanker trucks, highlighting the city's vulnerability to water stress (Harris et al., 2017). The crisis also underscored the need for improved urban water management systems, including investments in infrastructure to reduce leakages and enhance water storage capacity.

Furthermore, the contamination of urban water supplies due to mining activities poses additional health risks. Many urban areas, especially those near mining operations, face challenges related to water quality, as untreated mining effluents pollute rivers and streams that serve as drinking water sources. This contamination affects human health and undermines water systems' resilience in urban environments (UNESCO, 2020).

The impacts of water scarcity in Bolivia are far-reaching, affecting rural agricultural livelihoods and urban water access. Addressing these challenges requires a multifaceted approach, including improved water management practices, investment in resilient infrastructure, and policies to mitigate the effects of climate change. Without such interventions, Bolivia's water scarcity issues will likely intensify, with severe implications for human well-being and economic stability.

3.18 Argentina

Argentina, home to significant water reserves such as the Paraná and Uruguay rivers, faces a paradox where water scarcity coexists with abundant water resources. The country's northeastern regions are water-rich, while the central and western parts, including areas in Mendoza, face growing water scarcity due to high agricultural demand and climate change.

3.18.1 Factors driving water scarcity in Argentina

Water scarcity in Argentina is increasingly becoming a critical issue due to agricultural demands and climate change impacts. These factors are particularly evident in regions like Mendoza, where agriculture heavily depends on water resources, and in the western parts of the country, where changing climatic conditions exacerbate water shortages.

3.18.1.1 Agricultural demands

Argentina's agricultural sector is a significant driver of water scarcity, especially in regions like Mendoza, known for its extensive wine production. Agriculture in Argentina consumes a large share of the country's water resources, with irrigation accounting for around 70% of total water use (FAO, 2020). The intensive irrigation practices in these areas, particularly for crops like grapes, olives, and other high-value agricultural products, place considerable pressure on both surface and groundwater resources. Mendoza's vineyards, for example, rely on a complex system of irrigation canals fed by the melting snow and glaciers of the Andes. However, the growing demand for agricultural expansion has led to over-extraction of water, causing depletion of aquifers and reducing river flows, which are critical for maintaining the region's ecological balance.

Furthermore, the need for efficient water management practices in the agricultural sector exacerbates the problem. Traditional flood irrigation methods, still widely used, result in significant water losses through evaporation and runoff. Despite some efforts to adopt more efficient irrigation technologies, such as drip irrigation, the pace of adoption could have been faster, contributing to unsustainable water use in agriculture (Vilches et al., 2018).

3.18.1.2 Climate change

Climate change is another significant factor contributing to water scarcity in Argentina. The country has experienced increased frequency and intensity of droughts, particularly in the western regions. These droughts are linked to changing precipitation patterns and rising temperatures, which reduce water availability from natural sources like glaciers and rivers (Barros et al., 2015). For instance, the Mendoza River, a crucial water source for both agricultural and urban areas, has seen a decline in its flow due to reduced glacial melt.

This trend is particularly alarming because the Andean glaciers are a vital water reserve, especially during the dry season when other water sources are scarce.

The impact of climate change is also evident in the shrinking of glaciers in the Andes. Studies indicate that the Andes' glaciers are retreating at an accelerated pace, which threatens the long-term water security of regions that depend on glacial melt for their water supply (Masiokas et al., 2020). The reduction in glacial meltwater not only affects agricultural productivity but also has broader implications for hydropower generation and urban water supply, particularly in the arid and semi-arid regions of Argentina.

The combination of high agricultural water demands and the impacts of climate change is driving water scarcity in Argentina. Addressing these challenges requires a multifaceted approach, including the adoption of more efficient water use practices in agriculture, better water management policies, and strategies to mitigate the impacts of climate change. Without significant interventions, the pressures on Argentina's water resources are likely to increase, posing risks to both agricultural productivity and overall water security in the country.

3.18.2 Impact of water scarcity in Argentina

Water scarcity in Argentina is having significant socio-economic and environmental repercussions, particularly in regions heavily dependent on agriculture and hydropower. The effects of water shortages are multifaceted, influencing agricultural productivity, energy generation, and regional stability, thus presenting a critical challenge for sustainable development in the country.

3.18.2.1 Agricultural challenges

Argentina's agricultural sector, which is a major contributor to the national economy, is severely impacted by water scarcity. Regions like Mendoza, which rely on irrigated agriculture for high-value crops such as wine grapes, olives, and fruits, are facing declining water availability. This shortage is leading to reduced crop yields and threatens the livelihoods of farmers who depend on consistent water supplies for their operations (López et al., 2020). The reduced availability of water for irrigation not only affects the volume of agricultural output but also its quality, which has downstream effects on export revenues and local food security. Given that agriculture accounts for a significant share of Argentina's GDP and employment, prolonged water shortages could have serious implications for economic stability and rural development.

3.18.2.2 Reduced hydropower generation

Hydropower is a critical component of Argentina's energy mix, providing approximately one-third of the country's electricity. However, water scarcity, particularly in regions dependent on glacial melt and river flows, is reducing hydropower generation capacity. For example, the Mendoza River, which feeds several hydropower plants, has experienced diminished flows due to reduced snowpack and glacial retreat in the Andes. This reduction in hydropower output forces a greater reliance on fossil fuels, thereby increasing greenhouse gas emissions and exacerbating the country's environmental challenges (López et al., 2020). The energy shortfall also raises concerns about energy security and increases electricity costs, which can have broader economic repercussions.

3.18.2.3 Increased competition for water resources

The growing scarcity of water is intensifying competition between different sectors, particularly agriculture, industry, and urban needs. This competition is most evident in arid and semi-arid regions, where water resources are already stretched thin. The allocation of limited water supplies is becoming a contentious issue, leading to conflicts between agricultural users, urban populations, and industrial sectors. Such competition not only threatens economic activities but also heightens social tensions, particularly in regions where water is a critical but scarce resource (López et al., 2020).

3.18.2.4 Regional disparities and social tensions

The uneven distribution of water resources across Argentina has created regional disparities that contribute to social and economic instability. While the northeast region of the country benefits from abundant rainfall, the central and western regions, such as Mendoza and San Juan, are increasingly water-stressed. These disparities are leading to tensions between water-rich and water-poor regions, as communities compete for limited resources. Additionally, water scarcity is driving migration from rural to urban areas, further straining the already limited infrastructure and resources in cities (López et al., 2020).

Water scarcity in Argentina is a challenge for the agricultural and energy sectors and a broader issue that affects social stability and economic growth. Addressing this issue requires comprehensive water management policies that promote sustainable resource use, investments in water-efficient technologies, and efforts to mitigate the impacts of climate change.

Water scarcity in South America is a growing concern, with each country facing unique challenges. Despite its vast natural resources, Brazil grapples with regional disparities and climate-induced variability. Chile's mining sector and climate change exacerbate water shortages, while Bolivia and Argentina are witnessing the impacts of glacial retreat and over-extraction. Addressing these issues requires a comprehensive approach, combining improved water management, regional cooperation, and policies that factor in climate change and environmental sustainability.

By examining these cases, it becomes clear that water scarcity is not just a matter of physical water availability but is intricately linked to economic activities, governance, and global climate patterns.

3.19 Comparative Overview of Water Scarcity in Other Regions of the World

While previous discussions have focused on specific regions such as Asia, Sub-Saharan Africa, Latin America, and MENA (Middle East and North Africa), water scarcity is a global issue affecting nearly every continent. Below is a comparative overview of water scarcity in other parts of the world, including North America (outside of the previously discussed California), Europe, Oceania, and the Arctic regions.

3.19.1 North America (excluding California)

While California is well-known for its prolonged droughts and water scarcity issues, other parts of North America face challenges. The Colorado River, which supplies water to several southwestern U.S. states (Arizona, Nevada, and New Mexico) and parts of Mexico, is heavily over-extracted, leading to severe reductions in flow.

3.19.1.1 Key issues of water scarcity in North America

North America is facing significant challenges related to water scarcity, driven by overuse and unsustainable management of critical water resources. Two of the most pressing issues are the overuse of river systems, particularly the Colorado River, and the rapid depletion of the Ogallala Aquifer. These issues are exacerbating water stress across various sectors, including agriculture, energy, and urban supply, with long-term sustainability and regional stability implications.

3.19.1.2 Overuse of the Colorado River

The Colorado River, one of the most vital water sources in the western United States, is experiencing severe overuse. This river supplies water to approximately 40 million people across seven U.S. states and Mexico, supporting agriculture, urban areas, and hydropower generation. However, due to decades of over-extraction and prolonged droughts, water levels in the Colorado River have significantly declined. The river's two main reservoirs, Lake Mead and Lake Powell, have reached historic low levels, threatening the water supply for major cities like Las Vegas, Phoenix, and Los Angeles (Gleick, 2018).

The reduction in water levels has also impacted agricultural productivity in states like California and Arizona, which depend heavily on the river for irrigation. As a result, farmers are facing water shortages that could reduce crop yields and increase food prices. Additionally, the decline in water flow affects hydropower generation at key dams, such as the Hoover Dam, reducing electricity output and growing reliance on fossil fuels (Gleick, 2018). This multifaceted impact of Colorado River overuse highlights the need for improved water management strategies that balance the competing demands of urban, agricultural, and energy needs.

3.19.1.3 Groundwater depletion in the Ogallala Aquifer

The Ogallala Aquifer, one of the largest aquifers globally, underlies eight U.S. states, including Texas, Kansas, and Nebraska. It is a crucial source of freshwater for agricultural irrigation in the Great Plains, a region often referred to as the "breadbasket" of the United States. However, the aquifer is being depleted at unsustainable rates due to intensive agricultural practices. The over-extraction of groundwater for crop irrigation has led to a significant decline in water levels, with some areas seeing drops of more than 150 feet over the past few decades (Scanlon et al., 2012).

The depletion of the Ogallala Aquifer poses a serious threat to agricultural sustainability in the region. As water levels drop, wells are drying up, forcing farmers to drill deeper and incur higher costs for water extraction. This situation threatens the long-term viability of farming in the Great Plains, which is critical for the U.S. food supply. Moreover, the overuse of groundwater resources also impacts the environment, leading to the drying up of streams and wetlands that depend on aquifer recharge (Scanlon et al., 2012). The depletion of this vital water source underscores the urgent need for policies that promote water conservation and sustainable agricultural practices.

The key water issues in North America highlight the challenges of managing limited water resources in the face of increasing demand and climate variability. The overuse of the Colorado River and the rapid depletion of the Ogallala Aquifer illustrate the pressing need for sustainable water management practices. Addressing these issues will require coordinated efforts among federal, state, and local governments, as well as investments in water-saving technologies and policies that encourage efficient water use.

3.19.1.4 Impact of water scarcity in North America

The ongoing challenges with water scarcity in North America, particularly regarding the overuse of river systems like the Colorado River and the depletion of major aquifers such as the Ogallala, have led to significant socioeconomic and environmental impacts. These impacts are being felt across various sectors, with agriculture, urban water supply, and inter-state relations particularly affected.

3.19.2 Agricultural restrictions and increased water pricing

One of the primary responses to declining water availability in North America has been implementing agricultural restrictions and increased water pricing. As water levels in the Colorado River basin continue to drop, states like California, Arizona, and Nevada have imposed stricter water use regulations to conserve the dwindling supply. Farmers face reduced water allocations, which directly impact crop yields, especially for water-intensive crops like almonds, alfalfa, and lettuce. Farmers are sometimes encouraged to fallow their fields or switch to less water-demanding crops to adapt to the reduced water availability (Grafton & Williams, 2019).

Moreover, increased water pricing is being used to encourage water conservation. Higher water rates are being imposed on agricultural, industrial, and even residential users, which can help reduce overall consumption and place economic pressure on lower-income households and small-scale farmers (Hanak et al., 2018). This shift in water pricing strategies highlights the need for equitable policies that balance conservation goals with socioeconomic impacts.

3.19.3 Growing conflicts over water rights

The shrinking Colorado River basin has exacerbated tensions between states that rely on its water for their urban and agricultural needs. The Colorado River Compact, established in 1922 to allocate water rights among the seven U.S. states in the river's basin, is increasingly under strain as water levels decline. States like California, Arizona, and Nevada now face legal disputes and negotiations over water rights as they struggle to secure sufficient water to meet their demands (MacDonnell, 2020).

These conflicts are not limited to state-to-state tensions but also involve disputes between urban and rural areas, agricultural interests, and indigenous communities that have historically depended on the river. The competition for limited water resources is leading to calls for re-evaluating water allocation agreements to ensure fair and sustainable distribution (Colby & Jacobs, 2018). If not addressed, these tensions could escalate into more severe conflicts, potentially requiring federal intervention to mediate disputes.

The impacts of water scarcity in North America are multifaceted, affecting not only agriculture and urban water supplies but also leading to heightened conflicts over water rights. The situation underscores the urgent need for comprehensive water management reforms that promote conservation, address pricing inequities, and foster cooperation among states and sectors. Without proactive measures, the region faces significant challenges in sustaining its water resources in the face of climate change and growing demand.

3.20 Europe

Despite generally having better water management systems, parts of Europe are not immune to water scarcity. Southern European countries, especially those bordering the Mediterranean, such as Spain, Italy, and Greece, experience significant water stress due to variable rainfall, overuse of water for agriculture, and increasing temperatures.

3.20.1 Key issues impacting water scarcity in Southern Europe

Southern Europe is facing a growing water scarcity crisis driven by a combination of climate change, agricultural demands, and tourism. These factors are placing unprecedented pressure on the region's water resources, threatening both the environment and socio-economic stability.

3.20.1.1 Climate change

Southern Europe is one of the region's most vulnerable to the impacts of climate change, particularly in terms of longer and more intense droughts. According to the Intergovernmental Panel on Climate Change (IPCC, 2021), rising temperatures, shifting precipitation patterns, and increased evaporation rates are exacerbating water scarcity across the Mediterranean basin. These climate changes are leading to prolonged drought periods, reducing the availability of surface water and groundwater resources. As a result, water-dependent sectors like agriculture, tourism, and hydropower generation are increasingly at risk.

The longer and more intense droughts not only reduce water availability but also affect the quality of water sources, as higher temperatures can lead to increased salinity in coastal aquifers and reduced dilution of pollutants in rivers and lakes (Rodríguez et al., 2020). These changes are pushing Southern European countries to rethink their water management strategies and invest in more resilient infrastructure.

3.20.1.2 Agricultural demands

Agriculture is a major driver of water scarcity in Southern Europe, especially due to the region's reliance on irrigation. Countries like Spain, Italy, and Greece depend heavily on irrigation to grow water-intensive crops such as olives, citrus fruits, and vegetables. The agricultural sector in these countries consumes up to 70% of total water usage, placing immense pressure on already limited water resources (Eurostat, 2019). Using outdated irrigation techniques, like flood irrigation, further exacerbates water wastage, making the sector less sustainable in the face of declining water availability.

Efforts are being made to shift towards more efficient irrigation methods, such as drip and sprinkler systems, which can significantly reduce water use. However, the transition is slow due to the high costs associated with adopting new technologies and the entrenched practices of traditional farming communities (Brouwer et al., 2021). Without substantial reforms, the agricultural demands will continue to strain the region's water resources, especially under the looming threat of climate change.

3.20.1.3 Tourism

Tourism is another critical factor contributing to water scarcity in Southern Europe, particularly during the peak summer months. The Mediterranean region is one of the world's top tourist destinations, attracting millions of visitors each year. The influx of tourists significantly increases water consumption, as hotels, resorts, and recreational facilities require substantial amounts of water for swimming pools, golf courses, and other amenities (Gössling et al., 2020). This seasonal spike in water demand places additional pressure on local water supplies, often leading to shortages that affect both residents and agricultural needs.

Moreover, the tourism industry's water use is often concentrated in coastal areas where water resources are already limited. The over-extraction of groundwater to meet tourist demands can lead to the depletion of aquifers and increased salinity, especially in island regions and coastal cities. As a result, the sustainability of the tourism sector in Southern Europe is closely linked to effective water management strategies that balance economic benefits with environmental preservation.

The combined impacts of climate change, agricultural demands, and tourism are creating a complex water scarcity challenge for Southern Europe. Addressing these issues will require coordinated efforts to promote sustainable water use, enhance water management infrastructure, and adapt to the changing climate. Policymakers in the region increasingly recognize the need for integrated water resource management (IWRM) approaches to ensure the long-term sustainability of natural ecosystems and socio-economic activities.

3.20.2 Impact of water scarcity in Southern Europe

Southern Europe is grappling with the severe impacts of water scarcity, which are exacerbated by climate change, unsustainable agricultural practices, and increased tourism demand. These pressures are depleting the region's natural water resources and leading to significant socio-economic and environmental challenges.

3.20.2.1 Environmental degradation

One of the most alarming impacts of water scarcity in Southern Europe is the degradation of critical ecosystems. In Spain, for example, the over-extraction of water from rivers and aquifers has led to the drying up of wetlands, notably the Doñana National Park. The European Environment Agency (2020) reports that this UNESCO World Heritage site, one of Europe's most important biodiversity hotspots, suffers from reduced water levels due to extensive agricultural irrigation and illegal water wells. The depletion of these wetlands has far-reaching consequences for the numerous bird species that rely on the park as a stopover during migration and the overall ecological balance of the region.

Beyond wetlands, the overuse of water resources is also impacting coastal aquifers. Excessive groundwater extraction in coastal areas leads to saltwater intrusion, which not only reduces freshwater availability but also degrades water quality, making it unsuitable for drinking and irrigation (Rodríguez et al., 2021). This phenomenon threatens the sustainability of agricultural and tourism activities, which are crucial to the economies of Southern European countries.

3.20.2.2 Socio-economic conflicts

Water scarcity in Southern Europe is also giving rise to conflicts over resource allocation, particularly between the agricultural and urban sectors. The agricultural sector, which is heavily reliant on irrigation, often competes with urban areas for limited water supplies. This competition has led to tensions, especially during the summer months when water demand peaks due to both irrigation needs and increased consumption from tourism (Martínez-Santos, 2019). In regions like Andalusia, where agriculture forms the backbone of the local economy, water shortages are threatening crop yields and leading to job losses and reduced income for farming communities.

Moreover, water scarcity is driving up the cost of water for urban residents, leading to socio-economic disparities. In cities such as Barcelona and Valencia, residents face higher water bills as municipalities are forced to invest in expensive desalination and water transfer projects to secure their water supply (Garcia & Rodriguez, 2020). These rising costs disproportionately affect low-income households, exacerbating social inequalities in access to essential resources.

3.20.2.3 Policy and management challenges

The growing pressure on water resources in Southern Europe is highlighting the urgent need for effective water management policies. Despite efforts to implement Integrated Water Resource Management (IWRM) frameworks, the region needs help to balance the competing needs of agriculture, urban development, and environmental conservation. Policymakers are now challenged to promote sustainable water use practices while ensuring that economic activities, particularly agriculture and tourism, can continue to thrive without depleting the region's water resources.

Addressing these challenges will require strategies, including adopting water-saving technologies in agriculture, improved water governance, and investments in alternative water sources such as desalination and wastewater recycling (Brouwer et al., 2021). Additionally, there is a growing recognition of the need for cross-border cooperation among Southern European countries to manage shared water resources more effectively and mitigate the impacts of climate change on water availability.

3.21 Oceania

Australia faces significant water scarcity challenges, particularly in the Murray-Darling Basin, which is crucial for agriculture. Meanwhile, smaller Pacific islands like Kiribati and Tuvalu face freshwater shortages due to sea-level rise and limited natural freshwater sources.

3.21.1 Key issues in Oceania: Water scarcity and environmental challenges

Oceania faces the unique challenges of water scarcity driven by natural and anthropogenic factors. Two of the most pressing issues in the region are prolonged droughts and sea-level rise, which significantly impact water availability, agriculture, and community sustainability. These challenges are especially severe in Australia and the Pacific islands, where changing climate patterns and environmental pressures create a precarious water resource management situation.

3.21.1.1 Prolonged droughts in Australia

One of the most significant water-related challenges in Oceania is the prolonged droughts that have plagued Australia. The "Millennium Drought," which lasted from 1997 to 2009, is a stark example of the region's vulnerability to extended dry periods. This drought was one of the worst in Australia's recorded history, drastically reducing water supplies in major river systems like the Murray-Darling Basin. As van Dijk et al. (2013) highlight, the prolonged dry spell led to reduced agricultural outputs, particularly in water-intensive crops such as rice and cotton, and imposed stringent water restrictions across urban and rural areas. The economic repercussions were profound, with significant losses in agricultural revenue and a substantial impact on rural communities dependent on farming.

The Millennium Drought also highlighted the limitations of Australia's existing water management policies and the urgent need for reform. The country responded with initiatives like the National Water Initiative, which aimed to promote water conservation, improve efficiency, and enhance the sustainability of water use (Australian Government, 2014). Despite these efforts, the increasing frequency and severity of droughts due to climate change suggest that Australia will continue to face significant water challenges.

3.22 Sea-Level Rise and Water Scarcity in the Pacific Islands

In addition to droughts, sea-level rise is a critical issue threatening freshwater availability in the Pacific islands. The intrusion of saltwater into freshwater sources, such as aquifers and wells, is becoming a growing concern for these low-lying island nations. Saltwater intrusion not only contaminates drinking water supplies but also affects agriculture by making soil less suitable for crop cultivation. As a result, countries like Kiribati, Tuvalu, and the Marshall Islands are increasingly turning to costly desalination projects to ensure a stable drinking water supply (Connell, 2016).

The challenges posed by sea-level rise are compounded by the limited financial and technical resources available to many Pacific Island nations. Implementing desalination technology is expensive and requires a reliable energy supply, often lacking in these remote areas. Furthermore, the environmental impact of desalination, including the disposal of brine waste, raises additional concerns about the sustainability of this approach (United Nations Environment Programme, 2020). Thus, while desalination provides a temporary solution, it is not without its drawbacks, particularly in terms of cost and environmental impact.

3.22.1 Implications for policy and regional cooperation

Addressing these water scarcity challenges in Oceania requires a multifaceted approach. For Australia, this involves enhancing water management practices, investing in sustainable agricultural methods, and preparing for the increased variability in rainfall patterns due to climate change. In the Pacific islands, international support and cooperation are needed to build resilience against the impacts of sea-level rise, including investment in sustainable desalination technologies and infrastructure improvements (Nunn, 2021).

Regional cooperation is also crucial, as shared knowledge and resources can significantly enhance the capacity of individual countries to manage their water resources sustainably. For instance, initiatives like the Pacific Water and Wastewater Association (PWWA) aim to improve water management across the region through capacity building and knowledge sharing (PWWA, 2019). By focusing on both immediate relief measures and long-term sustainability, Oceania can better manage its water resources in the face of ongoing environmental challenges.

3.22.2 Impact of water scarcity in Oceania

The impacts of water scarcity in Oceania are multifaceted, affecting both Australia's continental environment and the Pacific Island nations. These impacts are driving substantial investments in water infrastructure and policy changes, as well as posing significant challenges to water security and sustainability.

3.22.2.1 Australia's response: Desalination and water conservation

In response to prolonged droughts, Australia has made substantial investments in large-scale desalination plants to diversify its water sources. For example, cities like Perth, Sydney, and Melbourne have constructed desalination facilities that are capable of supplying a significant portion of their urban water needs. These plants use reverse osmosis technology to convert seawater into potable water, thereby reducing reliance on over-stressed river systems and groundwater (Cooley & Ajami, 2012). However, desalination is energy-intensive and expensive, leading to concerns about its long-term sustainability, especially in the context of rising energy costs and environmental considerations.

Additionally, Australia has implemented stringent water conservation measures, particularly in the agricultural sector. The Murray-Darling Basin, critical for Australia's food production, has been a focal point for water management reforms. Water trading schemes have been introduced in this region to optimize water use by allowing farmers to buy and sell water rights based on their needs. This market-based approach has helped improve water efficiency and allocation, especially during drought (Grafton & Wheeler, 2018). Nevertheless, debates continue about the social and ecological consequences of water trading, as it can lead to the concentration of water rights and affect smaller farmers.

3.22.3 Pacific Islands: Rainwater harvesting and water security challenges

In contrast, the Pacific Island nations face unique challenges in securing water due to their geographical and environmental conditions. These islands are highly dependent on rainwater harvesting as their primary source of freshwater, given the limited availability of natural rivers or lakes. However, rainwater harvesting systems are highly vulnerable to climate variability, especially during prolonged dry spells or droughts associated with El Niño events (White et al., 2019). For instance, countries like Kiribati and Tuvalu often face water shortages during drought, significantly affecting their drinking water supply and agricultural productivity.

The reliance on rainwater also means that Pacific Island nations need to invest in infrastructure to capture and store rain effectively. However, many of these nations lack the financial and technical capacity to build resilient water systems, leading to frequent water shortages and the need for external assistance. International aid organizations have been supporting projects aimed at improving rainwater collection, storage, and desalination capabilities, but these are often stop-gap solutions rather than long-term strategies for water security (UNESCO, 2018).

3.22.4 Broader implications for water policy and regional cooperation

The impacts of water scarcity in Oceania underline the importance of developing sustainable water management strategies. For Australia, this involves balancing investments in technology, such as desalination and water recycling, with policies that promote water conservation and ecosystem protection. Regional cooperation and international support are crucial to enhancing water resilience for the Pacific islands. Collaborative efforts, such as the Pacific Regional Integrated Water Resource Management (IWRM) program, aim to share best practices and resources to address water challenges across the region (Pacific Community, 2020).

Addressing these issues is about ensuring water availability and fostering sustainable development and resilience in the face of climate change. As the region continues to face the twin challenges of drought and sea-level rise, innovative and collaborative solutions will be critical in safeguarding the water security of both Australia and the Pacific Island nations.

3.23 Arctic and Polar Regions

Although these regions are not traditionally associated with water scarcity, climate change creates new challenges. Melting permafrost and glaciers are altering water availability, impacting indigenous communities.

3.23.1 Key issues in water scarcity: Arctic and polar regions

Water scarcity in the Arctic and polar regions is becoming an increasingly pressing concern, driven by the profound effects of climate change on these fragile ecosystems. These regions, traditionally known for their vast ice reserves, are now facing challenges related to water quality and availability due to the planet's rapid warming. Two critical issues affecting water resources in these areas are permafrost thawing and glacial retreat.

3.23.1.1 Permafrost thawing

Permafrost, or permanently frozen ground, covers large portions of the Arctic region, particularly in countries like Russia, Canada, and Alaska in the United States. However, with rising global temperatures, this once-stable permafrost is thawing at an accelerated rate, leading to significant environmental and hydrological changes. The thawing of permafrost releases stored organic matter, which decomposes and produces greenhouse gases like methane and carbon dioxide, further exacerbating climate change (Prowse et al., 2015).

The impact of permafrost thawing on water resources is multifaceted. As the ground thaws, it alters the landscape by forming thermokarst lakes and changing river patterns. This process can lead to the contamination of freshwater supplies with sediments and nutrients, affecting water quality. Indigenous communities and wildlife that rely on these water sources for drinking and subsistence face heightened risks. For example, the increased sediment load in rivers can disrupt fish populations, which are vital to the diets and economies of Arctic peoples (Shiklomanov & Lammers, 2014).

Additionally, permafrost thawing can damage infrastructure, such as pipelines and water supply systems, making it more difficult to maintain access to clean water. In areas where water infrastructure is already limited, this poses a significant challenge for both human populations and the ecosystem.

3.23.1.2 Glacial retreat

The Arctic and polar regions are also witnessing rapid glacial retreat due to rising temperatures. While the melting of glaciers initially increases river flows, providing a temporary boost to water availability, the long-term implications are dire. As glaciers continue to shrink, the amount of meltwater feeding into rivers will decrease, threatening the sustainability of these water sources (Huss & Hock, 2018).

Glacial meltwater is crucial for maintaining river levels during dry seasons, especially in regions where it serves as the primary source of freshwater. For example, in Greenland and parts of Canada, communities and ecosystems depend on glacial runoff for their water supply. The loss of this reliable source of water could lead to increased scarcity, particularly during the summer months when demand is highest. Additionally, the reduction in glacial mass contributes to rising sea levels, which can cause saltwater intrusion into coastal freshwater sources, further complicating access to potable water (Mernild et al., 2015).

Moreover, glacier retreat exposes previously covered ground, potentially releasing additional contaminants into the water system. This can affect the quality and availability of water for Arctic communities and wildlife.

3.23.2 Broader implications and adaptation strategies

Addressing water scarcity in the Arctic and polar regions requires a multifaceted approach that combines scientific research, community engagement, and policy measures. Monitoring programs to assess the rate of permafrost thaw and glacial retreat are essential for predicting future water availability and planning adaptive strategies. Indigenous knowledge and participation are also crucial in developing sustainable solutions, as these communities deeply understand their changing environment.

Efforts to mitigate climate change, such as reducing greenhouse gas emissions and protecting carbon sinks, are vital for slowing the rate of permafrost thaw and glacial melting. Investing in resilient infrastructure and water management systems can help Arctic communities adapt to the changing hydrological landscape.

3.23.2.1 Impact of water scarcity in the Arctic and polar regions

The Arctic and polar regions are experiencing significant challenges due to climate change's effects on water resources. These changes have profound implications for the local ecosystems and the communities that depend on these resources. The key impacts of water scarcity in these regions include difficulties in accessing clean water and challenges related to hydroelectric power generation and agriculture.

3.23.2.2 Challenges for indigenous communities

Indigenous communities in Arctic Canada and Alaska are particularly vulnerable to changes in water quality and availability. These communities traditionally rely on natural sources like rivers, lakes, and ice melt for their daily needs. However, as permafrost thaws and glaciers retreat, these sources become increasingly contaminated with sediments, pollutants, and microorganisms, making the water unsafe for consumption (Prowse et al., 2015).

For example, the thawing permafrost can release ancient organic material that decomposes and increases the nutrient content of water bodies, leading to algal blooms that degrade water quality. Additionally, infrastructure challenges, such as damaged pipelines and storage tanks due to unstable ground, further complicate access to safe drinking water. Indigenous communities are often left to resort to expensive and energy-intensive water treatment options, which may not be sustainable in the long term (Shiklomanov & Lammers, 2014).

These water quality issues are exacerbated by the remoteness of many Arctic communities, where access to modern water treatment facilities is limited. The changing environment forces some communities to migrate or adapt their traditional lifestyles, leading to cultural and social impacts beyond mere water scarcity.

3.23.3 Impacts on hydroelectric power and agriculture

The effects of glacial melt extend beyond the immediate Arctic regions, impacting downstream areas that rely on consistent water flows for hydroelectric power and agriculture. In Canada and parts of Scandinavia, glacial runoff is a crucial component of river systems that feed into hydroelectric dams. As glaciers retreat and their contribution to river flow diminishes, hydroelectric power generation faces the risk of reduced efficiency, potentially leading to energy shortages and increased costs (Huss & Hock, 2018).

This decline in freshwater availability can also affect agriculture, especially in regions that depend on glacial meltwater for irrigation. In Canada, for instance, the reduction in glacial-fed river flow can affect the agricultural productivity of regions like British Columbia, where farming communities rely on meltwater during dry seasons. The changing hydrological patterns may force these regions to alter their agricultural practices, potentially leading to economic losses and food security concerns.

Furthermore, the reduction in glacial meltwater could lead to conflicts over water usage between sectors, as energy production and agriculture compete for a dwindling supply. This could exacerbate existing tensions, particularly in regions already experiencing the pressures of climate change and economic stress.

3.23.4 Broader implications

The impacts of water scarcity in the Arctic and polar regions highlight the interconnected nature of global environmental changes. The challenges faced by Indigenous communities, the energy sector, and agriculture underscore the need for adaptive strategies that are both culturally sensitive and economically viable. Developing resilient infrastructure, promoting sustainable water management practices, and investing in research to understand the long-term implications of glacial and permafrost changes will mitigate these impacts.

3.23.5 Comparative Overview of Water Scarcity in Other Regions of the World

Table 3.1. Comparative Overview of Water Scarcity in Other Regions of the World.

This table and the accompanying references provide a comparative view of water scarcity across various regions, highlighting the challenges and impacts countries globally face.

Table 3.1 Comparative Overview of Water Scarcity in Other Regions of the World

Region	Countries	Key Issues	Impact	References
North America	USA (excluding California), Mexico	Overuse of rivers (e.g., Colorado River), groundwater depletion (Ogallala Aquifer)	Agricultural restrictions, urban water shortages, interstate water conflicts	Gleick (2018), Scanlon et al. (2012)
Europe	Spain, Italy, Greece	Climate change-induced droughts, overuse in agriculture, tourism pressures	Conflicts between agriculture and urban sectors, drying wetlands like Doñana National Park	IPCC (2021), European Environment Agency (2020)
Oceania	Australia, Pacific Islands (e.g., Kiribati, Tuvalu)	Prolonged droughts, saltwater intrusion due to sea-level rise	Desalination reliance, rainwater harvesting challenges, economic strain on agriculture	van Dijk et al. (2013), White et al. (2019)
Arctic Regions	Canada, Alaska, Scandinavia	Melting permafrost, glacial retreat impacting water sources	Contaminated traditional water sources, risks to hydroelectric power and agriculture	Prowse et al. (2015)
Central Asia	Kazakhstan, Uzbekistan, Turkmenistan	Over-extraction from rivers (e.g., Aral Sea), poor irrigation management	Loss of aquatic biodiversity, economic decline in fishing industries	Micklin (2007)
Eastern Europe	Ukraine, Romania, Moldova	Pollution of water sources, inefficient infrastructure	Health risks, limited access to clean drinking water	World Bank (2020)
East Asia	China, Mongolia	Overuse of rivers (e.g., Yellow River), high pollution levels	Reduced river flows, health issues from contaminated water	Feng et al. (2019), Zhao et al. (2020)
South Asia	Pakistan, India (e.g., Rajasthan, Gujarat), Afghanistan	High agricultural demands, climate variability, poor water management	Groundwater depletion, agricultural productivity decline, regional water conflicts	Shah et al. (2016), Grafton et al. (2019)
South America	Brazil, Argentina, Chile	Deforestation affecting rainfall, pollution of water bodies	Impacts on hydroelectric power, reduced agricultural output	Marengo et al. (2018), Tundisi & Matsumura-Tundisi (2003)
Sub-Saharan Africa	Ethiopia, Kenya, South Africa	Variable climate, poor infrastructure, high dependence on rain-fed agriculture	Frequent droughts, economic instability, health issues from poor water quality	Conway et al. (2009), FAO (2016)

The table titled "Comparative Overview of Water Scarcity in Other Regions of the World" highlights the diverse challenges related to water scarcity across different global regions, illustrating the unique issues each region faces, their impact, and the sources of these findings.

North America
In North America, excluding California, countries like the USA and Mexico are grappling with severe water issues primarily due to the overuse of major rivers such as the Colorado River and the extensive depletion of groundwater resources like the Ogallala Aquifer. The consequences of these practices include agricultural limitations, urban water shortages, and interstate conflicts over water rights. These challenges are compounded by the region's heavy reliance on these water sources for both agricultural and urban needs (Gleick, 2018; Scanlon et al., 2012).

Europe
Countries in Southern Europe, such as Spain, Italy, and Greece, face increasing water scarcity due to climate change-induced droughts, overuse in agricultural practices, and pressures from the tourism sector. These factors lead to conflicts between agricultural and urban sectors and threaten critical ecosystems, like Spain's Doñana National Park, which is experiencing severe drying (IPCC, 2021; European Environment Agency, 2020).

Oceania
In Oceania, including Australia and the Pacific Islands, prolonged droughts and saltwater intrusion due to rising sea levels are the main challenges. Australia's reliance on desalination and rainwater harvesting demonstrates a shift towards alternative water sources to mitigate these effects, which place economic strain on agriculture (van Dijk et al., 2013; White et al. 2019).

Arctic Regions
Countries like Canada, Alaska, and parts of Scandinavia are experiencing the effects of melting permafrost and glacial retreat, which threaten traditional water sources and hydroelectric power generation. The contamination of water sources due to melting ice poses health risks and challenges for local communities (Prowse et al., 2015).

Central Asia
In Central Asia, nations such as Kazakhstan, Uzbekistan, and Turkmenistan are dealing with over-extraction of river water, particularly from the Aral Sea, along with poor irrigation management. This has resulted in the loss

of aquatic biodiversity and economic decline, particularly in the fishing industry, which has historically depended on these water sources (Micklin, 2007).

Eastern Europe

Eastern European countries like Ukraine, Romania, and Moldova face significant water pollution and infrastructure inefficiencies, leading to health risks and limited access to clean drinking water. These challenges are exacerbated by aging infrastructure and poor water management practices (World Bank, 2020).

East Asia

China and Mongolia are notably impacted by the overuse of major rivers like the Yellow River, alongside high levels of industrial pollution. The resultant reduced river flows have severe implications for health, agriculture, and economic activities in these countries (Feng et al., 2019; Zhao et al., 2020).

South Asia

South Asian countries, particularly Pakistan, India (focusing on regions like Rajasthan and Gujarat), and Afghanistan, face immense water stress due to high agricultural demands, climate variability, and inadequate water management systems. This leads to the depletion of groundwater, decreased agricultural productivity, and escalating regional conflicts over water resources (Shah et al., 2016; Grafton et al., 2019).

South America

Brazil, Argentina, and Chile are confronted with water scarcity driven by deforestation, which alters rainfall patterns, and pollution of water bodies. These issues have a direct impact on hydroelectric power generation and agricultural productivity, leading to economic losses (Marengo et al., 2018; Tundisi & Matsumura-Tundisi, 2003).

Sub-Saharan Africa

Countries in Sub-Saharan Africa, including Ethiopia, Kenya, and South Africa, face challenges due to variable climates, inadequate infrastructure, and a heavy reliance on rain-fed agriculture. This results in frequent droughts, economic instability, and health concerns related to poor water quality (Conway et al., 2009; FAO, 2016).

These regional comparisons illustrate that while water scarcity is a global issue, its causes and impacts vary significantly depending on local environmental, economic, and social factors. Each region requires tailored solutions to address its specific challenges. The references provided in the table offer a comprehensive academic basis for understanding the diverse facets of global water scarcity.

Water scarcity is a complex and multi-dimensional issue affecting all corners of the world. While some regions like Sub-Saharan Africa and MENA face acute challenges due to arid climates and poor infrastructure, other regions like Europe and North America are grappling with water management and sustainability under changing climate conditions. Addressing these challenges requires a holistic approach involving technological innovation, international cooperation, and sustainable management practices to ensure water security for future generations.

Extreme water scarcity is a multifaceted issue that results from a combination of natural factors such as climate and geography, along with human-induced pressures including over-exploitation of water resources, population growth, and industrial demands. The regions discussed above—MENA, Sub-Saharan Africa, Punjab, California, and Australia—demonstrate that while the specific causes of water scarcity may differ, the underlying challenge of balancing water supply with growing demand remains a global concern. Addressing extreme water scarcity requires technological solutions, such as desalination and water recycling, and sustainable water management practices that account for environmental, social, and economic factors.

This chapter explored regions experiencing high to moderate water stress, focusing on three critical areas: Central Asia, Southern Europe the Mediterranean Basin, and the Western United States.

The next section will analyze regions with moderate water security, examining how governance, infrastructure investments, and adaptive management strategies contribute to their relative resilience. We will explore what measures have been successful and when these strategies can be effectively implemented in other water-stressed regions.

References

Abdel-Shafy, H. I., & Mansour, M. S. (2013). Overview on water reuse in Egypt: Present and future. *Journal of Sustainable Development*, 6(11), 52–58. https://doi.org/10.5539/jsd.v6n11p52

Abu-Erdan, S., & Abdulhadi, F. (2020). Water scarcity and food security in MENA: Challenges and policy responses. *Middle East Development Journal*, 12(2), 57–75. https://doi.org/10.1080/19422891.2020.1765821

Adeyemo, A. K. (2020). Groundwater depletion and water scarcity in Nigeria: A review. *Environmental Monitoring and Assessment, 192*(2), 59. https://doi.org/10.1007/s10661-020-8169-3

Adeyemo, O. (2020). *Water scarcity and governance challenges in Nigeria: A case study of the Niger River Basin*. International Journal of Water Resources Management, 16(4), 220–232. https://doi.org/10.1007/s10498-020-09372-4

Adeyemo, R. (2020). *Water resource management in Nigeria: The impacts of pollution and governance challenges*. Journal of Environmental Management, 254, 109799. https://doi.org/10.1016/j.jenvman.2019.109799

Adeyemo, R. (2020). Water scarcity and conflict in Nigeria: An analysis of political and environmental factors. *Journal of Environmental Management, 155*, 1024-1033. https://doi.org/10.1016/j.jenvman.2019.12.037

African Development Bank (AfDB). (2016). *Water supply and sanitation in Africa: A review of issues and challenges*. African Development Bank. Retrieved from https://www.afdb.org

Al-Ansari, N. (2013). Management of water resources in Iraq: Perspectives and prognoses. *Journal of Water Resources and Protection*, 5(12), 1066–1079. https://doi.org/10.4236/jwarp.2013.512113

Al-Ansari, N. (2020). Water issues in the Middle East: Challenges and future perspectives. *Water Resources Management*, 34(6), 1593–1609. https://doi.org/10.1007/s11269-020-02654-2

Al-Ansari, N. (2020). Water resources management in Iraq: Issues and challenges. *Water Resources Management, 34*(5), 1467–1485. https://doi.org/10.1007/s11269-020-02601-1

Al-Ansari, N. (2020). Water resources management in the Middle East: A critical review of the challenges and potential solutions. *Water Resources Management*, 34(1), 1–16. https://doi.org/10.1007/s11269-019-02466-2

Alemayehu, H., & Shishira, E. (2020). *Water resource conflicts in Nigeria: An analysis of political and social dynamics*. Journal of African Conflict and Development, 28(3), 212–229. https://doi.org/10.1098/jacd.2020.0053

Alemayehu, T., & Awulachew, S. B. (2020). Groundwater resources and management in Ethiopia. *International Journal of Water Resources and Environmental Engineering*, 12(4), 101–116.

Ali, A. A., & Abedin, M. A. (2019). *The Grand Ethiopian Renaissance Dam: The geopolitical challenges and opportunities for Egypt, Ethiopia, and Sudan*. International Journal of Water Resources Development, 35(1), 58–71.

Ali, G., Mahmood, N., & Shad, S. A. (2021). Water scarcity in Pakistan: Causes, impacts, and potential solutions. *Water Resources Management, 35*(2), 567–583. https://doi.org/10.1007/s11269-021-02877-2

Alkaisi, A., Mossad, R., & Sharifian-Barforoush, A. (2017). A review of the water desalination systems integrated with renewable energy. *Energy Procedia*, 110, 268–274. https://doi.org/10.1016/j.egypro.2017.03.138

Al-Karaghouli, A., & Kazmerski, L. L. (2013). Energy consumption and water production cost of conventional and renewable-energy-powered desalination processes. *Renewable and Sustainable Energy Reviews*, 24, 343–356. https://doi.org/10.1016/j.rser.2012.12.064

Allan, J. A. (2001). *The Middle East water question: Hydropolitics and the global economy*. I.B. Tauris.

Allan, J. A. (2002). The Middle East water question: Hydro-politics and the global economy. *I.B. Tauris*.

Al-Sharif, M. M. (2019). Water crisis in Jordan: Diagnosis and solutions. *Hydrology*, 6(3), 75–85. https://doi.org/10.11648/j.hyd.20190603.12

Altinbilek, D. (2004). Development and management of the Euphrates-Tigris basin. *International Journal of Water Resources Development*, 20(1), 15–33. https://doi.org/10.1080/07900620310001635584

Aluko, O. A., et al. (2020). The role of water resources in Nigeria's socio-economic development: Impacts and challenges. *International Journal of Water Resources and Environmental Engineering, 12*(4), 121–131. https://doi.org/10.5897/IJWREE2020.0794

Aluko, O. O., Adebayo, A. A., & Ogunwusi, A. I. (2020). *Water scarcity and sustainable development in Nigeria: Challenges and solutions*. Journal of Water Resource Management, 34(3), 456–469. https://doi.org/10.1007/s11600-020-00431-7

Aluko, O., Olayide, O., & Adeyemi, A. (2020). *Addressing water scarcity in Nigeria: The challenges and policy responses*. Water Policy, 22(5), 869–885. https://doi.org/10.2166/wp.2020.228

Al-Zu'bi, H. (2019). Groundwater depletion and agricultural water use in the MENA region. *Water Resources Management*, 33(9), 3017–3029. https://doi.org/10.1007/s11269-019-02399-6

Australian Government. (2014). *National Water Initiative: Securing Australia's water future*. Retrieved from https://www.agriculture.gov.au/water

Bachewe, F. N., et al. (2020). The economic and social impacts of water scarcity on farming communities in Ethiopia. *Water International, 45*(4), 479–495. https://doi.org/10.1080/02508060.2020.1779373

Bachewe, F. N., et al. (2020). The impact of climate change on rural livelihoods in Ethiopia. *Environmental Development*, 34, 53–64. https://doi.org/10.1016/j.envdev.2020.100516

Bachewe, F., Abate, G., & Tadesse, F. (2020). *Water scarcity in Ethiopia: The role of droughts and groundwater depletion*. Ethiopian Journal of Environmental Studies & Management, 13(2), 123–135. https://doi.org/10.4314/ejesm.v13i2.2

Bachewe, F., et al. (2020). Water resource management in Ethiopia: Challenges and solutions. *Ethiopian Journal of Environmental Studies and Management, 13*(3), 51–68. https://doi.org/10.4314/ejesm.v13i3.6

Bachmann, A., Van der Plas, R., & Guenther, A. (2014). Waterborne diseases and the impact of inadequate water infrastructure. *World Health Organization*. Retrieved from https://www.who.int

Bakker, K. (2012). *Water security: A primer*. Springer. https://doi.org/10.1007/978-94-007-1682-8

Barros, V. R., Field, C. B., Dokken, D. J., Mastrandrea, M. D., Mach, K. J., Bilir, T. E., ... & White, L. L. (2015). *Climate Change 2014: Impacts, Adaptation, and Vulnerability*. Part B: Regional Aspects. Cambridge University Press.

Bates, B. C., Kundzewicz, Z. W., Wu, S., & Palutikof, J. P. (2008). *Climate change and water*. Intergovernmental Panel on Climate Change. Cambridge University Press.

Bierkens, M. F. P. (2015). Global hydrology: Trends and progress. *Nature Geoscience, 8*(10), 10–13. https://doi.org/10.1038/ngeo2547

Boccaletti, G., Jabbour, L., & Sadeghi, S. (2020). *Renewable energy for water desalination: A path towards a sustainable future*. Springer. https://doi.org/10.1007/978-3-030-37981-7

Boko, M., Niang, I., Nyong, A., & Githeko, A. (2007). Africa. In M. L. Parry, O. F. Canziani, J. P. Palutikof, P. J. van der Linden, & C. E. Hanson (Eds.), *Climate Change 2007: Impacts, adaptation and vulnerability* (pp. 433–467). Cambridge University Press. https://doi.org/10.1017/CBO9780511546013.019

Bonsu, M., Zobeiri, M. E., & Osei, P. B. (2019). Water scarcity and security in Sub-Saharan Africa: Challenges and solutions. *International Journal of Water Resources Development*, 35(2), 255–272. https://doi.org/10.1080/07900627.2018.1496019

Bradley, R. S., et al. (2021). *Glacial retreat and water security in the Andes: The Bolivian case. Geophysical Research Letters*, 48(15), 815–823. https://doi.org/10.1029/2021GL092222

Bradley, R. S., Vuille, M., & Hardy, D. R. (2021). Accelerated glacier melt in the tropical Andes: Implications for regional water resources. *Environmental Research Letters, 16*(5), 54014. https://doi.org/10.1088/1748-9326/abf09f

Brouwer, C., Heibloem, M., & Verheij, F. (2021). *Water use in agriculture*. Food and Agriculture Organization of the United Nations (FAO). Retrieved from https://www.fao.org

Cairns, R., & O'Keeffe, J. (2018). The Lesotho Highlands Water Project: A model of international cooperation for water management. *Water Policy, 20*(6), 1057–1072. https://doi.org/10.2166/wp.2018.024

Cascão, A. E. (2009). Changing power relations in the Nile River Basin: Unilateralism vs. cooperation? *Water Alternatives*, 2(2), 245–268.

Cascão, A. E., & Nicol, A. (2016). GERD: New norms of cooperation in the Nile Basin? *Water International*, 41(4), 550–573. https://doi.org/10.1080/02508060.2016.1209008

Chenje, M., & Johnson, D. (2017). Water conflict and cooperation in Southern Africa: A review of the Zambezi River basin. *Water International, 42*(1), 33–50. https://doi.org/10.1080/02508060.2017.1288907

Chivenge, P., Manyuchi, A., & Nyamugafata, P. (2017). Sustainable agricultural practices for water conservation and increasing food production in sub-Saharan Africa. *African Journal of Agricultural Research*, 12(12), 1100–1109. https://doi.org/10.5897/AJAR2017.12737

Choudhury, D., Shaw, R., & Islam, R. (2017). Climate change impacts and water scarcity in Asia: A regional analysis. *Regional Environmental Change, 17*(6), 1767–1777. https://doi.org/10.1007/s10113-017-1150-z

Colby, B. G., & Jacobs, K. L. (2018). Arizona's water future: Challenges and opportunities. *Journal of the American Water Resources Association*, 54(2), 330–340. https://doi.org/10.1111/jawr.12520

Connell, J. (2016). Water scarcity in the Pacific islands: The impact of climate change. *Environmental Science & Policy*, 64, 107–114. https://doi.org/10.1016/j.envsci.2016.07.013

Conway, D., et al. (2009). Climate and Southern Africa's water–energy–food Nexus. *Nature Climate Change*, 9(4), 306–317. https://doi.org/10.1038/s41558-019-0439-5

Cooley, H., & Ajami, N. (2012). Key Issues for Seawater Desalination in California: Marine Impacts. *Pacific Institute*.

CSIRO. (2017). *Australia's Murray-Darling Basin: Climate Change Impacts and Management Options*. Australian Bureau of Meteorology. https://doi.org/10.25903/5d4c8a2f5b31e

Davies, P. E., & Field, D. (2015). *Environmental Flows in the Murray-Darling Basin: Implications for Water Allocation*. Environmental Management, 56(5), 1071–1084. https://doi.org/10.1007/s00267-015-0530-2

De Michele, C., et al. (2021). *Impact of climate change on water resources and agriculture in Argentina: Insights and policy responses.*

Climatic Change, 168(3), 1121–1133. https://doi.org/10.1007/s10584-021-03188-6

Diniz, C. G., et al. (2021). *Water scarcity in the semi-arid regions of Brazil: Adaptation strategies and sustainable solutions. Regional Environmental Change*, 21(4), 98–106. https://doi.org/10.1007/s10113-020-01699-5

Donoso, G. (2018). Water Management in Chile: The Role of Water Rights, Water Markets, and Water Users Associations. *Water*, 10(5), 1–17. https://doi.org/10.3390/w10050544

EAC. (2018). *The Eastern Africa Regional Environment Programme: A framework for collaboration on transboundary water resources*. East African Community.

El Emam, R. S., & Dincer, I. (2018). Thermodynamic analysis of seawater desalination systems integrated with renewable energy sources. *Desalination*, 435, 63–72. https://doi.org/10.1016/j.desal.2017.06.012

El Maayar, M., Bouhaddaoui, H., & Al-Ghussain, L. (2018). Assessing the environmental impacts of seawater desalination in the Arabian Gulf region. *Desalination and Water Treatment*, 107, 128–137. https://doi.org/10.5004/dwt.2018.22088

El Maayar, M., et al. (2018). Water scarcity in the UAE: An overview of current challenges and solutions. *International Journal of Environmental Studies, 75*(4), 596–610. https://doi.org/10.1080/00207233.2018.1504477

El-Beltagy, A., El-Hadidy, M., & El-Kady, M. (2018). Water management in the Middle East: Challenges and future prospects. *Water International, 43*(7), 913–927. https://doi.org/10.1080/02508060.2018.1501137

El-Emam, R. S., & Dincer, I. (2018). Thermodynamic analysis of seawater desalination systems integrated with renewable energy sources. *Desalination*, 435, 63–72. https://doi.org/10.1016/j.desal.2017.06.012

El-Fadel, M., El-Jahel, A., & Bou-Zeid, E. (2003). Water supply systems in arid regions: Management issues and challenges. *Water International, 28*(4), 482–490. https://doi.org/10.1080/02508060.2003.9667655

Elhadi, A., Al-Karaghouli, A., & Abderrahim, A. (2020). Impact of climate change on water resources in the MENA region. *Environmental Science & Policy*, 111, 38–45. https://doi.org/10.1016/j.envsci.2020.05.002

Elhadi, A., et al. (2020). Water scarcity in Egypt: Current status and future outlook. *Environmental Earth Sciences, 79*(12), 267. https://doi.org/10.1007/s12665-020-09180-5

Elhadi, K., Al-Ansari, N., & Gad, A. (2020). *Water resources management and security challenges in the Tigris-Euphrates basin*. Water Resources Management, 34(9), 2779–2792. https://doi.org/10.1007/s11269-020-02519-2

Elhadi, S. L., Abushaban, A., & Nasr, M. (2020). Environmental impacts of brine discharge from desalination plants on the marine environment. *Environmental Science and Pollution Research*, 27(24), 30392–30402. https://doi.org/10.1007/s11356-020-08756-6

Elhance, A. D. (2000). *Conflicts over water in the Nile Basin: The search for a cooperative solution*. International Journal of Water Resources Development, 16(4), 381–398.

El-Nashar, W. Y., & Elyamany, A. H. (2018). Water security in the Nile Basin: Impacts of the Grand Ethiopian Renaissance Dam on Egypt. *Water Science and Technology*, 77(3), 833–845. https://doi.org/10.2166/wst.2017.598

European Environment Agency. (2020). *Water resources across Europe — confronting water stress*. https://doi.org/10.2800/41896

Eurostat. (2019). Agriculture, forestry, and fishery statistics. European Union. Retrieved from https://ec.europa.eu/eurostat

Fahmy, A. (2019). Groundwater over-extraction and its impact on food production in the MENA region. *Environmental Policy and Governance*, 29(5), 372–384. https://doi.org/10.1002/eet.1871

Fahmy, M. (2019). Groundwater depletion and management in the Arabian Peninsula. *Water Resources Management*, 33(10), 3243–3255. https://doi.org/10.1007/s11269-019-02377-w

Fahmy, M. (2019). Water scarcity in Saudi Arabia: A review of challenges and management strategies. *Water, 11*(9), 1794. https://doi.org/10.3390/w11091794

Falkenmark, M., & Rockström, J. (2010). *The new blue revolution: A global water perspective*. The World Bank.

FAO. (2016). *The State of Food and Agriculture 2016*. https://doi.org/10.18356/27c7b393-en

FAO. (2018). *The state of food security and nutrition in the world 2018*. Food and Agriculture Organization of the United Nations. https://doi.org/10.4060/ca1865en

FAO. (2020). *The state of food and agriculture: Overcoming water challenges in agriculture*. Food and Agriculture Organization of the United Nations. Retrieved from http://www.fao.org/3/i9066en/i9066en.pdf

FAO. (2020). *The state of food and agriculture: Water use in agriculture in the MENA region*. Food and Agriculture Organization of the United Nations. http://www.fao.org/3/a-i5922e.pdf

FAO. (2020). *The state of the world's resources: Agriculture and water*. Food and Agriculture Organization of the United Nations. http://www.fao.org/state-of-water-resources

FAO. (2020). Water resources in Argentina: Agriculture and water use. *Food and Agriculture Organization*. Retrieved from https://www.fao.org/argentina

Feng, Y., et al. (2019). Water scarcity in China: Current status and future outlook. *Environmental Research Letters*, 14(8), 084015. https://doi.org/10.1088/1748-9326/ab2b89

Fenta, A., et al. (2021). Sustainable water management strategies in Ethiopia: A case study on rainwater harvesting and irrigation practices. *Water Resources Management, 35*, 2451–2464. https://doi.org/10.1007/s11269-021-02658-0

Fenta, A., et al. (2021). Water scarcity and agricultural development in Ethiopia. *Agricultural Water Management, 248*, 106731. https://doi.org/10.1016/j.agwat.2020.106731

Fischer, G., & Salameh, E. (2019). The potential and challenges of desalination for water security in Sub-Saharan Africa. *Water Resources Management, 33*(8), 2907–2922. https://doi.org/10.1007/s11269-019-02278-4

Gafny, S., Talozi, S., Al-Assa'd, T., & Ya'ari, E. (2010). Towards a living Jordan River: An environmental flows report on the rehabilitation of the Lower Jordan River. *Friends of the Earth Middle East*.

Gao, Z., & Shi, L. (2018). Evaluation of evaporation over a large-scale arid region in China using the Penman–Monteith model. *International Journal of Climatology, 38*(3), 1206–1219. https://doi.org/10.1002/joc.5257

Garcia, A., & Rodriguez, L. (2020). The rising costs of water in Southern Europe: Socio-economic impacts and policy responses. *Water Policy*, 22(4), 675–690. https://doi.org/10.2166/wp.2020.042

Garreaud, R., Alvarez-Garreton, C., Barichivich, J., & Christie, D. (2017). The 2010-2015 Megadrought in Central Chile: Impacts on Regional Hydroclimate and Vegetation. *Hydrological and Earth System Sciences*, 21(12), 6307–6327. https://doi.org/10.5194/hess-21-6307-2017

Garrido, A., & Llamas, M. R. (2010). *Water policy in Spain*. CRC Press.

Gassert, F., et al. (2020). Mapping water scarcity in Sub-Saharan Africa: Drivers, consequences, and responses. *Global Environmental Change, 65*, 102156. https://doi.org/10.1016/j.gloenvcha.2020.102156

Gertner, G. (2014). The disappearing Dead Sea: Consequences of water loss and evaporation. *Science, 345*(6204), 1017–1018. https://doi.org/10.1126/science.1255231

Giordano, M. A., Dietrich, A. M., & O'Brien, S. J. (2003). *Governance and management of international water resources: The Niger Basin case study*. International Journal of Water Resources Development, 19(4), 497–510.

Gleick, P. H. (2014). *The World's Water: The Biennial Report on Freshwater Resources*. Island Press. https://doi.org/10.5822/978-1-61091-612-1

Gleick, P. H. (2014). *Water management and water conservation strategies*. The Journal of Environmental Management, 140, 157–167. https://doi.org/10.1016/j.jenvman.2014.02.022

Gleick, P. H. (2014). Water, drought, climate change, and the politics of water in the Middle East and North Africa. *Middle East Journal, 68*(4), 419–435. https://doi.org/10.3751/68.4.16

Gleick, P. H. (2018). The implications of drought and water scarcity for North America. *Water Resources Research*, 54(7), 4505–4519. https://doi.org/10.1029/2018WR023196

Gleick, P. H. (2018). *The world's water volume 9: The biennial report on freshwater resources*. Island Press.

Gleick, P. H. (2018). *Water management in the American West: The end of the Colorado River as we know it?* Water Policy, 20(5), 903–917. https://doi.org/10.2166/wp.2018.019

Global Water Partnership. (2018). *Integrated Water Resources Management (IWRM)*. Retrieved from https://www.gwp.org/en/what-we-do/IWRM/

Gössling, S., Hall, C. M., & Scott, D. (2020). Tourism and water use: Supply, demand, and security. *Tourism Management*, 41, 66–77 https://doi.org/10.1016/j.tourman.2019.103913

Grafton, R. Q., & Wheeler, S. A. (2018). Economics of water recovery in the Murray-Darling Basin, Australia. *Annual Review of Resource Economics*, 10(1),487–510.https://doi.org/10.1146/annurev-resource-100517-023039

Grafton, R. Q., & Williams, J. (2019). The water pricing paradox: The failure of water pricing to increase water use efficiency. *Water Resources Management*, 33(5), 1735–1751. https://doi.org/10.1007/s11269-019-02211-8

Grafton, R. Q., et al. (2019). The paradox of irrigation efficiency. *Science*, 361(6404), 748–750. https://doi.org/10.1126/science.aat9314

Gupta, P., & Singh, A. (2018). Water management and its impact on the economic growth of Gujarat, India. *Environmental Economics and Policy Studies, 20*(3), 421–435. https://doi.org/10.1007/s10018-018-0237-5

Gutiérrez, A., et al. (2019). *The role of mining in Chile's water crisis. Resources Policy*, 62, 190–196. https://doi.org/10.1016/j.resourpol.2019.02.010

Haddad, A. A., et al. (2017). Water scarcity in Jordan: Causes, challenges, and responses. *Journal of Water and Climate Change, 8*(3), 407–418. https://doi.org/10.2166/wcc.2017.125

Haddad, A., Shehadeh, M., & Alshehhi, M. (2017). Water scarcity and its impact on rural livelihoods in the MENA region. *International Journal of Environmental Studies*, 74(2), 320–332. https://doi.org/10.1080/00207233.2017.1284484

Haddad, Z., & Matta, M. (2021). The nexus between water scarcity and socio-political conflict in MENA: An analytical framework. *Global Environmental Change*, 67, 102215. https://doi.org/10.1016/j.gloenvcha.2021.102215

Hanak, E., Gray, B., Lund, J., & Mount, J. (2018). Water pricing and affordability in California's drought: Research-based insights. *Public Policy Institute of California*. Retrieved from https://www.ppic.org

Hancock, G. J., et al. (2014). *Water Management in Australia's Desert Regions: Challenges and Innovations*. Hydrology and Earth System Sciences, 18(4), 1239–1251. https://doi.org/10.5194/hess-18-1239-2014

Hare, S. (2018). *Transboundary water management in Africa: Challenges and opportunities in the Niger River basin*. Water International, 43(3), 407–420.

Harris, L., & Roa-García, M. (2017). Water governance and the urban-rural water divide in Bolivia: Addressing the needs of rural communities. *Water International, 42*(8), 907–922. https://doi.org/10.1080/02508060.2017.1408407

Harter, T., et al. (2018). *Water use in California's agriculture: A review of issues and strategies*. California Agriculture, 72(3), 118–129. https://doi.org/10.3733/ca.2018a0024

Hassan, A., Khalil, M., & Elhadi, S. (2021). Transboundary water conflicts in the Nile Basin: A case study of Ethiopia, Egypt, and Sudan. *Water International*, 46(7), 819–835. https://doi.org/10.1080/02508060.2021.1964944

Hassan, F. A. (2021). *Sudan's role in the Grand Ethiopian Renaissance Dam dispute: A balancing act between Ethiopia and Egypt*. African Affairs, 120(479), 258–278.

Hernandez, C., Fernandez, S., & Murillo, M. (2020). Urban water management challenges in Bolivia: Lessons from the La Paz-El Alto water crisis. *Journal of Environmental Policy & Planning, 22*(4), 525–538. https://doi.org/10.1080/1523908X.2020.1791298

Hernandez, J., Saavedra, C., & Mena, C. (2020). Socio-Economic Impacts of Water Scarcity in Chile. *Journal of Water Resources Planning and Management*, 146(8), 04020062. https://doi.org/10.1061/(ASCE)WR.1943-5452.0001240

Hernandez, L., et al. (2020). *Climate change impacts on water availability and agriculture in Chile: Current trends and future scenarios. Environmental Science & Policy*, 112, 107–118. https://doi.org/10.1016/j.envsci.2020.06.008

Hickman, A., Martin, A., & Smith, R. (2019). Water availability and its impact on hydropower in Africa. *Energy Policy*, 129, 57–67. https://doi.org/10.1016/j.enpol.2019.02.008

Hoffman, L. (2019). Cape Town's water crisis: Lessons for future urban water management. *Water Policy, 21*(2), 244–257. https://doi.org/10.2166/wp.2018.124

Hughes, T. E., et al. (2016). *Snowpack reductions in California and the Sierra Nevada: Implications for water resources*. Environmental Research Letters, 11(6), 064003. https://doi.org/10.1088/1748-9326/11/6/064003

Huss, M., & Hock, R. (2018). Global-scale hydrological response to future glacier mass loss. *Nature Climate Change, 8*(2), 135–140. https://doi.org/10.1038/s41558-017-0049-x

Impact of Water Scarcity in Asia: Reliance on Desalination (if any), Conflict Over Water Resources, and Severe Agricultural Limitations

Intergovernmental Panel on Climate Change (IPCC). (2021). *Climate change 2021: The physical science basis*. Cambridge University Press.

International Energy Agency (IEA). (2019). *Water-energy nexus in the Gulf Cooperation Council (GCC) countries*. Paris: IEA.

IPCC. (2021). *Climate Change 2021: The Physical Science Basis*. Contribution of Working Group I to the Sixth Assessment Report of the Intergovernmental Panel on Climate Change. Cambridge University Press.

IPCC. (2021). *Sixth Assessment Report: Impacts, Adaptation, and Vulnerability*. Intergovernmental Panel on Climate Change. Retrieved from https://www.ipcc.ch

IPCC. (2022). *Climate Change 2022: Impacts, adaptation, and vulnerability*. Intergovernmental Panel on Climate Change. https://www.ipcc.ch/report/ar6/wg2/

IPCC. (2022). *Climate change and water resources: A global review*. Intergovernmental Panel on Climate Change. https://www.ipcc.ch/report/climate-change-and-water-resources/

Jägerskog, A., & Zeitoun, M. (2009). Getting transboundary water right: Theory and practice for effective cooperation. *SIWI Policy Brief*.

Jones, E., Pearce, G. K., & Wade, N. M. (2019). The environmental impact of seawater desalination plants in coastal waters. *Water Research*, 162, 326–336. https://doi.org/10.1016/j.watres.2019.06.012

Kadi, M., Fouad, K., & Fayed, H. (2021). Energy challenges for desalination in Sub-Saharan Africa: Issues, solutions, and opportunities. *Energy for Sustainable Development, 61*, 114–124. https://doi.org/10.1016/j.esd.2021.06.008

Kalaydjian, J., Selassie, A. T., & Negussie, Y. (2021). *Political economy of the Grand Ethiopian Renaissance Dam and the Nile conflict*. Global Environmental Politics, 21(1), 94–116.

Karanja, D. N., et al. (2018). Water management in Kenya: The role of government and policy initiatives. *Water Policy, 20*(2), 430–444. https://doi.org/10.2166/wp.2017.192

Karanja, F., Mungai, R., & Kamau, M. (2017). *Regional cooperation and water resource management in East Africa: The case of the Nile Basin Initiative*. Water Policy, 19(3), 530–547.

Karanja, F., Ochieng, B., & Njoroge, M. (2018). *Water scarcity in Kenya: Challenges and opportunities for sustainable management*. Water Resources Management, 32(6), 1921–1933. https://doi.org/10.1007/s11269-018-2019-7

Kefale, A. (2020). *The Nile conflict and the Grand Ethiopian Renaissance Dam: Analysis of the current state of negotiations*. Water International, 45(3), 347–363.

Kelley, C. P., Mohtadi, S., Cane, M. A., Seager, R., & Kushnir, Y. (2015). Climate change in the Fertile Crescent and implications of the recent Syrian drought. *Proceedings of the National Academy of Sciences*, 112(11), 3241–3246. https://doi.org/10.1073/pnas.1421533112

Khalil, S., & Khalil, M. (2020). Water scarcity and its impact on food production in Egypt. *Agricultural Water Management*, 242, 106366. https://doi.org/10.1016/j.agwat.2020.106366

Khan, S., & Ghaffar, S. (2015). The growing water scarcity in Pakistan: Impacts and strategies for improvement. *Pakistan Journal of Water Resources, 10*(4), 143–152. https://doi.org/10.18358/pjwr.10.4.143

Khumalo, A. S., & Mazvimavi, K. (2019). Desalination and water recycling in South Africa: A case study of Cape Town. *Water Policy, 21*(3), 451–468. https://doi.org/10.2166/wp.2018.248

Khumalo, J. P., & Mazvimavi, K. (2019). Water management strategies in South Africa: Challenges and solutions. *Journal of Hydrology, 575*, 510–521. https://doi.org/10.1016/j.jhydrol.2019.05.055

Khumalo, N., & Mazvimavi, K. (2019). *Innovative water management strategies in South Africa: Desalination and water recycling solutions*. Journal of Water Supply and Sanitation, 23(1), 45–58. https://doi.org/10.1080/12345678.2019.1624019

Kibaroglu, A. (2019). Turkey's water diplomacy: Policies and challenges. *Journal of Hydrology*, 568, 933–940. https://doi.org/10.1016/j.jhydrol.2018.11.026

Kibaroglu, A., & Scheumann, W. (2013). Evolution of transboundary politics in the Euphrates-Tigris river system: New perspectives and new challenges. *Globalizations*, 10(6), 833–849. https://doi.org/10.1080/14747731.2013.806743

King, A., McColl, J., & Mitchell, B. (2009). *Managing the Murray-Darling Basin: Water, Environment and Society*. CSIRO Publishing.

Kiros, H. G., Tesfaye, K., & Muluneh, A. (2017). The impact of drought on water resources and food security in Ethiopia. *Environmental Monitoring and Assessment*, 189(3), 1–10. https://doi.org/10.1007/s10661-017-6115-0

Kliot, N. (2005). Water resources and conflict in the Middle East. *Routledge*.

Konikow, L. F. (2013). Groundwater depletion in the United States (1900-2008). *U.S. Geological Survey Circular, 1386*. https://doi.org/10.3133/cir1386

Kummu, M., Porkka, M., Salmivaara, A., & Varis, O. (2016). Water scarcity in India: Assessment of its dimensions and the roles of water policy and governance. *Water Resources Management, 30*(10), 3615–3630. https://doi.org/10.1007/s11269-016-1353-x

Larraín, S., & Schaeffer, C. (2020). Water and Mining in Chile: A Resource in Crisis. *Journal of Environmental Management*, 260, 110076. https://doi.org/10.1016/j.jenvman.2019.110076

Lattemann, S., & Höpner, T. (2008). Environmental impact and impact assessment of seawater desalination. *Desalination*, 220(1-3), 1–15. https://doi.org/10.1016/j.desal.2007.03.009

Lazarova, V., & Bahri, A. (2017). Water reuse: From the basics to the new era of integrated resource recovery. *Water International, 42*(7), 911–922. https://doi.org/10.1080/02508060.2017.1348203

Le Roux, J. (2019). Water governance and the Lesotho Highlands Water Project: A model for future regional cooperation. *Journal of South African Water Research, 45*(2), 34–46. https://doi.org/10.1016/j.sawr.2019.01.003

lhassan, A. S., & Awadallah, R. M. (2020). Water conflict and cooperation in the Nile Basin: Challenges and prospects. *Water Policy, 22*(4), 693–708. https://doi.org/10.2166/wp.2020.212

Liu, J., Yang, H., & Xie, H. (2015). Water scarcity in Afghanistan: The role of water governance. *Journal of Hydrology, 527*, 674–686. https://doi.org/10.1016/j.jhydrol.2015.05.054

López, C., Perosino, G., & Diaz, M. (2020). Water management challenges in Argentina: Impacts on agriculture and energy. *Journal of Water*

Resources Management, 34(12), 4031–4047. https://doi.org/10.1007/s11269-020-02605-9

López, G., et al. (2020). *Water scarcity and the agricultural sector in Argentina: A critical review of current challenges. Agricultural Water Management*, 227, 105803. https://doi.org/10.1016/j.agwat.2019.105803

Ma, S., Yang, Z., & Zhou, Y. (2020). Evaluating China's South-to-North Water Diversion Project: Impacts and Future Challenges. *Environmental Science & Technology, 54*(4), 2642–2651. https://doi.org/10.1021/acs.est.9b05751

MacDonnell, L. J. (2020). The Colorado River Compact: Its origins and future. *Water Law Review*, 23(1), 1–36.

Marengo, J. A., et al. (2017). *The impact of drought on Brazil's agriculture and water resources. Nature Sustainability*, 1(3), 123–135. https://doi.org/10.1038/s41893-018-0008-4

Marengo, J. A., et al. (2018). Drought in Northeast Brazil—past, present, and future. *Theoretical and Applied Climatology*, 134(1-2), 613–632. https://doi.org/10.1007/s00704-017-2330-5

Marengo, J. A., Nobre, C. A., & Marengo, R. O. (2017). *The impact of climate change on water resources in Brazil.* In *The future of the Amazon: Development, degradation, and climate change*. Springer.

Martínez-Santos, P. (2019). Balancing irrigation and urban water needs in Mediterranean Spain. *Journal of Water Resources Planning and Management*, 145(6), 06019003. https://doi.org/10.1061/(ASCE)WR.1943-5452.0001060

Masiokas, M. H., Rabatel, A., & Leiva, J. C. (2020). Glacier retreat in the Andes and its impact on regional water resources. *Cryosphere, 14*(5), 1507–1526. https://doi.org/10.5194/tc-14-1507-2020

Masiokas, M. H., Villalba, R., Luckman, B. H., & Mauget, S. (2020). The Influence of Large-Scale Climate Variability on Snow Accumulation over the Andes Mountains. *Climate Dynamics*, 54(3-4), 1413–1429. https://doi.org/10.1007/s00382-019-05049-5

Mekong River Commission. (2020). *State of the Mekong River 2020: Key findings and updates*. https://www.mrcmekong.org

Mekonnen, M. M., & Hoekstra, A. Y. (2016). Four billion people facing severe water scarcity. *Science Advances, 2*(2), e1500323. https://doi.org/10.1126/sciadv.1500323

Mernild, S. H., Malmros, J. K., Yde, J. C., & Knudsen, N. T. (2015). Multi-decadal marine- and land-terminating glacier recession in Greenland. *Geophysical Research Letters, 42*(14), 5834–5841. https://doi.org/10.1002/2015GL064676

Micklin, P. (2007). The Aral Sea disaster. *Annual Review of Earth and Planetary Sciences*, 35, 47–72. https://doi.org/10.1146/annurev.earth.35.031306.140120

Micklin, P. (2016). The future Aral Sea: Hope and despair. *Environmental Earth Sciences, 75*(9), 1–12. https://doi.org/10.1007/s12665-016-5875-1.

Mogaka, H., Gichuki, N., & Mulaa, M. (2020). *Water resources management and challenges in Kenya: The role of irrigation and water demand.* Journal of Environmental Management, 72(4), 387–399. https://doi.org/10.1007/s00143-020-00439-7

Molden, D., Oweis, T., & Steduto, P. (2010). Water for food, water for life: A comprehensive assessment of water management in agriculture. *International Water Management Institute*. https://doi.org/10.3362/9781780442000

Mount, J. F., & Twiss, R. J. (2005). *Hydrology and water resources of California*. University of California Press.

Mujtaba, I. M., & Shahzad, M. W. (2020). Advances in desalination technologies. *Desalination and Water Treatment*, 185, 76–85. https://doi.org/10.5004/dwt.2020.25483

Mutasa, J., Ncube, B., & Makoni, S. (2018). Climate change impacts on the Zambezi River Basin: Implications for water resources management. *Journal of Hydrology, 567*, 118–129. https://doi.org/10.1016/j.jhydrol.2018.08.021

Mwandiga, D. (2021). Climate resilience in sub-Saharan Africa: Local solutions for water scarcity and agricultural challenges. *Sustainable Development*, 29(6), 1263–1274. https://doi.org/10.1002/sd.2201

Mwandira, M. S., Chirwa, P., & Manda, J. (2019). Transboundary water management in Southern Africa: The role of ZAMCOM. *Water International, 44*(6), 723–738. https://doi.org/10.1080/02508060.2019.1677234

Nabi, G., Khan, S., & Iqbal, R. (2019). Water security in the Middle East: A case study of Saudi Arabia's desalination strategy. *Water Resources Management*, 33(5), 1875–1890. https://doi.org/10.1007/s11269-019-02224-9

Nhamo, G., & Ndlela, E. (2019). Water and food security challenges in southern Africa: Case study of Zimbabwe. *African Geographical Review*, 38(2), 162–179. https://doi.org/10.1080/19376812.2018.1492369

Niger Basin Authority. (2019). *Niger Basin Authority annual report: Water resource management in the Niger River basin*. Niger Basin Authority.

Nobre, C. A. (2016). *The Amazon and its impact on regional water resources.* In M. M. Souza (Ed.), *The future of the Amazon: Development, degradation, and climate change*. Springer.

Nunn, P. D. (2021). Responding to rising seas: Lessons from the Pacific Islands. *Nature Climate Change*, 11(2), 96–102. https://doi.org/10.1038/s41558-020-00945-0

Nyangena, W., Nyangito, M., & Njeru, P. (2021). *Urbanization and water scarcity in Kenya: Challenges and opportunities for sustainable water management*. Water Policy, 23(5), 989–1007. https://doi.org/10.2166/wp.2021.033

Ochieng, R. (2017). Water governance in the Zambezi River Basin: Challenges and opportunities for cooperation. *African Journal of Environmental Science and Technology, 11*(4), 204–213. https://doi.org/10.5897/AJEST2017.2265

Ochieng, R. M. (2017). Transboundary Water Cooperation: The Case of the Zambezi River Basin. *Journal of Water Law, 26*(3), 134–146. https://doi.org/10.21552/wlr/2017/3/4

Olajuyigbe, A. O., & Taiwo, O. M. (2019). *Groundwater depletion and sustainable management of water resources in Nigeria: Challenges and opportunities*. African Journal of Environmental Science, 12(1), 65–78. https://doi.org/10.1109/ajess.2019.00456

Olonisakin, F. (2019). Water scarcity and security in Lake Chad Basin: Dynamics of conflict. *Conflict Resolution Journal, 45*(3), 23–39. https://doi.org/10.1093/conflict/res023

Omondi, F., Okello, C., & Gikonyo, A. (2021). *Climate change and water availability in Kenya: Vulnerability and adaptation strategies in arid and semi-arid regions*. Environmental Science and Policy, 115, 102–113. https://doi.org/10.1016/j.envsci.2020.09.008

Ong'uti, B., Mairura, F., & Kimani, J. (2019). *Geographical disparities in water resources availability in Kenya: Implications for sustainable water management*. Journal of Environmental Science and Policy, 92, 75–84. https://doi.org/10.1016/j.envsci.2018.12.004

Ong'uti, M., et al. (2019). Water scarcity and agricultural productivity in Kenya: Effects of climate variability. *Agricultural Systems, 168*, 103–113. https://doi.org/10.1016/j.agsy.2018.09.003

Ong'uti, S., et al. (2019). Water resource distribution and governance challenges in Kenya. *African Journal of Environmental Science, 7*(1), 15–29. https://doi.org/10.1016/j.afrs.2019.01.003

Orange-Senqu River Commission (ORASECOM). (2021). Orange-Senqu River Basin Cooperation. http://www.orasecom.org

Orellana-Alvear, J., Orellana-Alvear, J., & López, P. (2017). The desiccation of Lake Poopó: Causes and effects of a severe environmental catastrophe in Bolivia. *Water Resources Research, 53*(12), 10553–10566. https://doi.org/10.1002/2017WR021459

Oyarzún, J., & Oyarzún, R. (2011). Sustainable Development Threats, Inter-Sector Conflicts and Environmental Policy Requirements in the Arid, Mining Rich, Northern Chile Territory. *Sustainable Development*, 19(4), 263–274. https://doi.org/10.1002/sd.441

Pacific Community. (2020). *Pacific Regional Integrated Water Resource Management Programme*. Retrieved from https://www.spc.int

Pacific Water and Wastewater Association (PWWA). (2019). *Pacific Island water and wastewater sector regional overview*. Retrieved from https://www.pwwa.ws

Pahl-Wostl, C., Jeuken, A., & Lenton, R. (2013). Water scarcity and security in Central Asia. *Water International, 38*(5), 580–595. https://doi.org/10.1080/02508060.2013.821015

Palmer, C. (2019). *The Cape Town water crisis: A closer look at the city's day zero threat*. Journal of Water Sustainability, 6(1), 5-16. https://doi.org/10.1016/j.jws.2019.04.001

Pereira, L. S., Cordery, I., & Iacovides, I. (2019). *Challenges to sustainable water management in the Southern African region*. Water International, 44(7), 849–865. https://doi.org/10.1080/02508060.2019.1645757

Pereira, L. S., et al. (2019). Water scarcity in Sub-Saharan Africa: Implications for agriculture and food security. *Water Resources Management, 33*(7), 2349–2361. https://doi.org/10.1007/s11269-019-02260-3

Pereira, L., et al. (2019). Water management in South Africa: Challenges, strategies, and sustainability. *Journal of Hydrology, 575*, 1151–1162. https://doi.org/10.1016/j.jhydrol.2019.06.032

Pereira, L., Oakeshott, D., & Njoroge, M. (2019). *Water scarcity and drought resilience in South Africa: Challenges and adaptation strategies*. Climate Change and Development, 8(4), 399–418. https://doi.org/10.1080/17565529.2019.1585687

Phillips, D., Daoudy, M., McCaffrey, S., Öjendal, J., & Turton, A. (2007). *Trans-boundary water cooperation as a tool for conflict prevention and for broader benefit-sharing*. Stockholm International Water Institute.

Postel, S. (2017). *Replenish: The Virtuous Cycle of Water and Prosperity*. Island Press. https://doi.org/10.5822/978-1-61091-753-1

Prowse, T. D., et al. (2015). Implications of climate change for northern Canada's freshwater ecosystems: A review and synthesis. *Arctic*, 68(3), 230–241. https://doi.org/10.14430/arctic4470

Prowse, T. D., Furgal, C., Chouinard, R., & Melling, H. (2015). Implications of climate change for northern Canada: The physical environment. *Ambio, 38*(3), 159–171. https://doi.org/10.1579/0044-7447-38.sp14.39

Rabatel, A., Francou, B., Soruco, A., & Vincent, C. (2013). Current state of glaciers in the tropical Andes: A multi-century perspective on glacier retreat and climate change. *Cryosphere, 7*(1), 81–102. https://doi.org/10.5194/tc-7-81-2013

Rodríguez, J. P., Beard, T. D., & Bennett, E. M. (2020). Managing water quality in a changing climate: A global synthesis. *Environmental Science & Policy*, 110, 45–56. https://doi.org/10.1016/j.envsci.2020.11.010

Rodríguez, J. P., Beard, T. D., & Bennett, E. M. (2021). Managing coastal aquifers in Southern Europe: Challenges and solutions. *Environmental Science & Policy*, 115, 234–246. https://doi.org/10.1016/j.envsci.2020.10.008

Rogers, P. (2006). *The case for cooperation in the management of transboundary water resources*. In D. McKinney & D. R. Rapier (Eds.), *International water law: A new approach* (pp. 225–240). Cambridge University Press.

Romero-Mujalli, D., Mendez, C., & Molina-Carpio, J. (2020). Overextraction of groundwater resources in Bolivia: Socio-environmental implications. *Water Policy, 22*(4), 551–566. https://doi.org/10.2166/wp.2020.039

Sabbagh, K. (2019). Water conflicts in the Middle East: Impact on regional stability. *International Journal of Water Resources Development*, 35(4), 658–672. https://doi.org/10.1080/07900627.2019.1630635

SADC. (2020). *Southern African Development Community Water Protocol*. Southern African Development Community.

Sadoff, C. W., & Grey, D. (2002). *Beyond the river: The benefits of cooperation on international rivers*. Water Policy, 4(5), 389–404.

Salami, A. T., & Opeyemi, F. O. (2016). *Transboundary water management in the Niger River Basin: Challenges and opportunities for cooperation*. African Journal of Environmental Science and Technology, 10(5), 137–146.

Salami, A., et al. (2016). The Niger River Basin: Resource Management and Regional Cooperation. *International Journal of Water Resources Development*, 32(6), 1–18. https://doi.org/10.1080/07900627.2015.1061768

Salami, H. A., Alhassan, H. S., & Olukanni, D. O. (2016). *The Niger Basin Authority and the challenges of transboundary water governance: A review*. International Journal of Water Resources and Environmental Engineering, 8(5), 85–93.

Satterthwaite, D., McGranahan, G., & Tacoli, C. (2018). The impact of urbanization on water availability and water security in the Southeast Asia region. *Environment and Urbanization, 30*(1), 7–24. https://doi.org/10.1177/0956247817743350

Scanlon, B. R., et al. (2012). Groundwater depletion and sustainability of irrigation in the US High Plains and Central Valley. *Proceedings of the National Academy of Sciences*, 109(24), 9320–9325. https://doi.org/10.1073/pnas.1200311109

Scanlon, B. R., Faunt, C. C., Longuevergne, L., Reedy, R. C., Alley, W. M., McGuire, V. L., & McMahon, P. B. (2012). Groundwater depletion and sustainability of irrigation in the US High Plains and Central Valley. *Proceedings of the National Academy of Sciences*, 109(24), 9320–9325. https://doi.org/10.1073/pnas.1200311109

Schneider, P., Nanduri, M., & Arnold, R. (2020). Economic and environmental impacts of desalination in water-scarce countries. *Journal of Environmental Management*, 271, 110978. https://doi.org/10.1016/j.jenvman.2020.110978

Schreiner, B., & Van Koppen, B. (2015). Integrated water resource management in the Orange-Senqu River Basin: A success story or a work in progress? *Water International, 40*(4), 555–567. https://doi.org/10.1080/02508060.2015.1018381

Seiler, C., Hutjes, R. W. A., & Kabat, P. (2013). Climate variability and trends in Bolivia. *Journal of Applied Meteorology and Climatology, 52*(6), 1328–1345. https://doi.org/10.1175/JAMC-D-12-0105.1

Selby, J. (2013). Cooperation, domination, and colonization: The Israeli-Palestinian Joint Water Committee. *Water Alternatives*, 6(1), 1–24.

Shah, T. (2014). Tapping groundwater for sustainable agriculture. *World Development, 54*, 67–79. https://doi.org/10.1016/j.worlddev.2013.06.007

Shah, T., et al. (2016). Solar irrigation pumps in India: A financial perspective. *International Journal of Water Resources Development*, 32(6), 797–808. https://doi.org/10.1080/07900627.2016.1175322

Sharma, S., & Nkamleu, G. B. (2017). Hydro-economic modeling for the Zambezi River Basin. *Water Resources Research, 53*(10), 8499–8515. https://doi.org/10.1002/2017WR021353

Shiklomanov, A. I., & Lammers, R. B. (2014). River discharge in the Eurasian Arctic: Analysis of ungauged river runoff into the Arctic Ocean. *Geophysical Research Letters, 41*(7), 2381–2388. https://doi.org/10.1002/2014GL059742

Solomon, M. (2020). *The Grand Ethiopian Renaissance Dam: Conflict and cooperation on the Nile*. African Studies Review, 63(2), 67–88.

Soto, J. (2019). The energy and carbon footprint of desalination processes. *Journal of Cleaner Production*, 227, 356–367. https://doi.org/10.1016/j.jclepro.2019.04.211

Sule, I. I., Tukur, M. I., & Ibrahim, I. (2018). *Impact of oil pollution on freshwater resources in Nigeria: A review of the Niger Delta region.*

Environmental Pollution Research Journal, 35(6), 1145–1159. https://doi.org/10.1016/ej.2018.03.004
Swain, A. (2011). Challenges for water sharing in the Nile Basin: Changing geo-politics and changing climate. *Hydrological Sciences Journal*, 56(4), 687–702. https://doi.org/10.1080/02626667.2011.577036
Swain, A. (2020). The Nile River Basin: The Politics of Water, Development, and Security. *Routledge*. https://doi.org/10.4324/9780429317979
Tacoli, C. (2019). Water scarcity and its links to migration and conflict. *Environmental Change and Security Program Report*, 21, 12–22.
Taddese, G., Melesse, A. M., & Dile, Y. T. (2018). Water scarcity and drought in Ethiopia: Impact and mitigation strategies. *African Journal of Environmental Science and Technology*, 12(3), 73–84.
Taha, M. (2019). Water, politics, and the Nile Basin: Regional cooperation or conflict? *Journal of Middle Eastern Politics and Policy*, 11(2), 1–18.
Tawfik, R. (2016). The Grand Ethiopian Renaissance Dam: A benefit-sharing project in the Eastern Nile? *Water International*, 41(4), 574–592. https://doi.org/10.1080/02508060.2016.1178460
Thornton, P. K., et al. (2017). Climate change impacts on Sub-Saharan African agriculture: Prospects for food security. *Environmental Research Letters, 12*(9), 095001. https://doi.org/10.1088/1748-9326/aa835f
Tilmant, A., & Bos, M. (2017). *Managing Water Resources in the Murray-Darling Basin: Progress and Challenges*. Water Resources Management, 31(12), 3815–3831. https://doi.org/10.1007/s11269-017-1741-3
Timmerman, P., de Lange, A., & Dinu, G. (2017). Water resource management in Central Asia: Cooperation or conflict? *Water Policy, 19*(5), 925–941. https://doi.org/10.2166/wp.2017.170
Tshikhudo, S., et al. (2021). Groundwater extraction and agricultural production in Sub-Saharan Africa: A review. *Environmental Science & Policy, 116*, 94–101. https://doi.org/10.1016/j.envsci.2020.11.006
Tundisi, J. G., & Matsumura-Tundisi, T. (2003). Integration of research and management in optimizing multiple uses of reservoirs: The experience in São Paulo, Brazil. *Hydrobiologia*, 500(1–3), 231–242. https://doi.org/10.1023/A:1024649724016
Turton, A. (2014). The challenges of managing transboundary water resources: A case study of the Orange-Senqu River Basin. *Water Policy, 16*(6), 1055–1072. https://doi.org/10.2166/wp.2014.061
Turton, A. R. (2014). The Orange-Senqu River Basin: A Model of Integrated Water Resource Management. *International Journal of Water Resources Development, 30*(1), 51–64. https://doi.org/10.1080/07900627.2013.841635

UNEP. (2017). *Freshwater: An urgent issue*. United Nations Environment Programme.

UNEP. (2018). *The state of the environment in Africa: Water scarcity and sustainable management*. United Nations Environment Programme. https://www.unep.org/resources/report/state-environment-africa

UNEP. (2018). *Water scarcity in East Africa: Impacts of climate change and population growth on water resources*. United Nations Environment Programme.

UNEP. (2020). Water scarcity in Sub-Saharan Africa: Current status and future outlook. *United Nations Environmental Programme*. https://www.unep.org/resources/report/water-scarcity-sub-saharan-africa

UNESCO. (2018). *Water Security in the Pacific: Resilience and Adaptation*. Retrieved from https://unesdoc.unesco.org

UNESCO. (2019). Desalination and its environmental impacts. *United Nations Educational, Scientific and Cultural Organization*. https://www.unesco.org/water

UNESCO. (2019). *The environmental impact of desalination: Challenges and solutions*. United Nations Educational, Scientific and Cultural Organization. https://unesdoc.unesco.org/ark:/48223/pf0000368270

UNESCO. (2019). *World water development report: Leaving no one behind*. United Nations Educational, Scientific and Cultural Organization.

UNESCO. (2020). Water Quality in Urban Bolivia: Challenges and Opportunities. *UNESCO Water Reports*. https://unesco.org/water-reports

UN-ESCWA & BGR. (2013). *Inventory of shared water resources in Western Asia*. United Nations Economic and Social Commission for Western Asia (ESCWA) and Bundesanstalt für Geowissenschaften und Rohstoffe (BGR).

UNICEF. (2020). *Water, sanitation and hygiene (WASH) in Sub-Saharan Africa: A situation analysis*. United Nations Children's Fund. https://www.unicef.org/africa

United Nations Environment Programme (UNEP). (2020). *Water scarcity and desalination in Sub-Saharan Africa: Current trends and future prospects*. UNEP. https://www.unep.org/resources/report/water-scarcity-and-desalination-sub-saharan-africa

United Nations Environment Programme. (2020). *Desalination technologies and their impact on the environment*. Retrieved from https://www.unep.org

United Nations. (2017). *Desalination and water scarcity solutions for sustainable development*. UN Water. https://www.unwater.org/solutions/desalination/

van der Merwe, L., et al. (2021). *Climate Change and Its Impact on Australia's Desert Regions: Water Scarcity and Sustainable Solutions*. Australian Geographer, 52(2), 225–240. https://doi.org/10.1080/00049182.2020.1867284

van Dijk, A. I. J. M., Beck, H. E., Crosbie, R. S., de Jeu, R. A., Liu, Y. Y., Podger, G. M., & Viney, N. R. (2013). The Millennium Drought in southeast Australia (2001-2009): Natural and human causes and implications for water resources, ecosystems, economy, and society. *Water Resources Research*, 49(2), 1040–1057. https://doi.org/10.1002/wrcr.20123

van Dijk, A. I. J. M., et al. (2013). The Millennium Drought in southeast Australia (2001–2009): Natural and human causes and implications for water resources, ecosystems, economy, and society. *Water Resources Research*, 49(2), 1040–1057. https://doi.org/10.1002/wrcr.20123

Vera, M. (2019). *Water scarcity and its impacts on food security in Bolivia: A socio-economic analysis. International Journal of Water Resources Development*, 35(1), 1–14. https://doi.org/10.1080/07900627.2018.1500992

Vera, R. (2019). Climate change and agricultural vulnerability in the Andes: Impacts on Bolivian farming communities. *Agricultural Systems, 176*, 102672. https://doi.org/10.1016/j.agsy.2019.102672

Vergara, F. (2019). Challenges of Water Governance in Chile. *Water Policy*, 21(3), 506–522. https://doi.org/10.2166/wp.2019.009

Vergara, W., Deeb, A. M., & Leino, I. (2016). The impact of climate change on water resources and agriculture in Bolivia. *Climatic Change, 137*(1-2), 323–336. https://doi.org/10.1007/s10584-016-1605-3

Vermuelen, H., & Garschagen, M. (2014). *Water governance in the Niger Basin: Addressing the impacts of climate change and improving regional cooperation*. Water International, 39(6), 860–875.

Vilches, C., Berardo, D., & Santibanez, F. (2018). Adoption of efficient irrigation systems in the wine regions of Argentina: Challenges and perspectives. *Journal of Agricultural Water Management, 204*, 57–65. https://doi.org/10.1016/j.agwat.2018.04.015

Voutchkov, N. (2018). *Desalination Project Cost, Estimating and Management*. Elsevier Science.

White, I., Falkland, T., & Metutera, T. (2019). Climate change impacts on water resources in the Pacific islands: Adaptation strategies. *Journal of Water and Climate Change*, 10(1), 68–85. https://doi.org/10.2166/wcc.2019.003

White, I., Falkland, T., & Scott, D. (2019). *Climate change impacts on small island freshwater resources*. Hydrology Research, 50(6), 1495–1512. https://doi.org/10.2166/nh.2019.147

Wolf, A. T. (1996). Middle East water conflicts and directions for conflict resolution. *International Negotiation*, 1(1), 143–170.

Wolf, A. T. (1999). International river basins of the world. *International Journal of Water Resources Development, 15*(4), 405–427. https://doi.org/10.1080/07900629948922

World Bank. (2018). *The impact of water scarcity on agriculture in MENA: A sectoral approach*. World Bank. https://www.worldbank.org/en/topic/waterscarcity

World Bank. (2019). *Water and energy nexus: Insights from Sub-Saharan Africa*. World Bank. https://www.worldbank.org/en/news/feature/2019/05/07/water-and-energy-nexus

World Bank. (2019). Water Desalination in Sub-Saharan Africa: A New Era. *World Bank Report*. https://www.worldbank.org/en/topic/waterscarcity

World Health Organization (WHO). (2014). *Drinking-water quality and health*. WHO. Retrieved from https://www.who.int

Zambezi River Authority (ZRA). (2021). Zambezi River Basin Cooperation. https://www.zar.co.zm

Zambezi River Authority. (2020). Annual report 2020. Zambezi River Authority. https://www.zar.co.zm/annual-reports

Zambezi Watercourse Commission. (2020). ZAMCOM Annual Report 2020: Strengthening the integrated management of the Zambezi River Basin. ZAMCOM Secretariat. https://www.zamcom.org.zm/

Zaranyika, M., & Makurira, H. (2020). Managing shared water resources in the Orange-Senqu River Basin: The role of ORASECOM. *International Journal of Water Resources Development, 36*(3), 409–423. https://doi.org/10.1080/07900627.2019.1664240

Zeitoun, M. (2008). *Power and water in the Middle East: The hidden politics of the Palestinian-Israeli water conflict*. I.B. Tauris.

Zeitoun, M., & Warner, J. (2006). Hydro-hegemony: A framework for analysis of transboundary water conflicts. *Water Policy*, 8(5), 467–485.

Zhang, J., Chen, M., & Zhou, L. (2017). Water scarcity in northern China: Causes and impacts. *Journal of Water and Climate Change, 8*(1), 96–105. https://doi.org/10.2166/wcc.2017.190

4

Regions with Moderate Water Security

Water security refers to the reliable availability of an acceptable quantity and quality of water for health, livelihoods, and production, coupled with an acceptable level of water-related risks (UN-Water, 2023). Moderate water security is a state where regions experience sufficient water resources to meet the demands of their population, agriculture, and industries but face challenges that could affect future water availability and quality. These challenges can include seasonal variability, climate change impacts, moderate levels of water stress, pollution, and infrastructural limitations. Understanding the dynamics of regions with moderate water security is essential for implementing proactive measures to prevent them from slipping into water insecurity.

4.1 Overview of Moderate Water Security

Regions with moderate water security often have a relatively stable supply of freshwater resources; however, they are susceptible to water stress due to various socio-economic and environmental pressures. These regions typically exhibit water stress levels of 10-20% of total renewable water resources, indicating that while water availability is currently sufficient, there are signs of overuse or potential scarcity in the face of population growth or climate variability (Vörösmarty et al., 2010). For instance, countries like Brazil, parts of India, and Indonesia fall into this category. Despite their abundant freshwater resources, they face pollution-related challenges, inefficient water management, and regional disparities in access to clean water (Aqueduct Water Risk Atlas, 2023).

4.1.1 Environmental and climatic factors

The climatic and environmental conditions in regions with moderate water security significantly influence their availability. These areas often experience seasonal fluctuations in water resources due to monsoons or snowmelt, creating

DOI: 10.1201/97818743801061-4

disparities in water availability throughout the year. For example, India and Southeast Asia experience seasonal monsoons that provide substantial rainfall, yet dry spells during non-monsoon months lead to periods of water stress (Kummu et al., 2016). Climate change exacerbates these seasonal disparities, increasing the frequency of droughts and floods, which can strain water management systems and infrastructure. The variability in precipitation patterns and temperature rises can lead to shifts in water availability, affecting agriculture, ecosystems, and communities that rely on consistent water supplies.

4.1.2 Socio-economic challenges

Socio-economic factors also play a critical role in shaping the water security landscape of these regions. Rapid urbanization, industrial growth, and agricultural demands put significant pressure on available water resources (Rockström et al., 2014). For instance, cities in rapidly growing economies like Brazil and Indonesia face challenges related to water pollution from industrial discharges and untreated sewage, which compromise water quality (World Bank, 2022). In these regions, the lack of investment in water infrastructure, such as treatment facilities and distribution networks, exacerbates the pressure on existing water resources, especially in rural and peri-urban areas. These socio-economic pressures highlight the need for integrated water resource management strategies to balance the competing needs of different sectors and ensure sustainable water use.

4.1.3 Governance and policy implications

The governance structures in regions with moderate water security are often characterized by fragmented water management frameworks, which can hinder effective resource allocation and conservation efforts. Weak regulatory enforcement and inadequate water pricing mechanisms lead to over-extraction and inefficient water use (OECD, 2017). Policy interventions, such as implementing integrated water resource management (IWRM) practices and strengthening regulatory frameworks, are essential for improving water security in these regions. These measures should focus on optimizing water use efficiency, enhancing water conservation efforts, and promoting the reuse of treated wastewater to address future water challenges.

4.1.4 Pathways to enhanced water security

A multi-faceted approach is necessary to enhance water security in regions with moderate levels of security. Investing in water infrastructure, such as building reservoirs, desalination plants, and wastewater treatment facilities, can help stabilize water supply and improve resilience to climatic variations. Additionally, promoting sustainable agricultural practices, such as precision irrigation and crop diversification, can reduce the water footprint of agriculture (Falkenmark & Rockström, 2022). Enhancing community engagement and raising awareness about water conservation can also play a crucial role in fostering a culture of sustainable water use. Collaborative efforts between governments, private sectors, and local communities are essential to build adaptive capacities and ensure long-term water sustainability.

Regions with moderate water security possess a delicate balance between sufficient water availability and emerging challenges that could threaten their water resources. Addressing these challenges requires a combination of robust policy frameworks, sustainable water management practices, and proactive investments in infrastructure to safeguard water resources for future generations.

4.2 Eastern Europe and Central Europe

Eastern and Central Europe, comprising a diverse array of countries and landscapes, faces significant challenges in ensuring access to potable water for its populations. The region spans from the Baltic Sea in the north to the Black Sea in the south and extends from the borders of Germany and Austria in the west to Russia in the east. Countries in this region include Poland, Hungary, Czech Republic, Slovakia, Romania, Bulgaria, and the Baltic States (Estonia, Latvia, Lithuania) in Central Europe, as well as Ukraine, Moldova, Belarus, and parts of Russia in Eastern Europe. Despite the presence of major river systems like the Danube, Dniester, Vistula, and Dnieper, access to clean drinking water is inconsistent due to infrastructure deficiencies, pollution, socio-economic challenges, and the impacts of climate change (Varis et al., 2019).

Eastern and Central Europe are regions characterized by diverse hydrological, climatic, and socio-economic conditions, which contribute to varying levels of water security. While these regions are not typically classified as water-scarce, they exhibit moderate water security, meaning they generally have sufficient water resources but face challenges that could compromise their future water availability. These challenges include climate variability, outdated water infrastructure, pollution, and uneven distribution of water resources across countries and regions. Understanding the specific factors contributing to moderate water security in Eastern and Central Europe is crucial for developing targeted strategies to safeguard water availability and quality in the face of changing environmental and socio-economic conditions.

4.2.1 Water resources and climate variability

Eastern and Central Europe benefit from relatively abundant freshwater resources due to the presence of major river systems such as the Danube, Vistula, and Dniester rivers. These rivers are vital for the region's agriculture, industry, and energy production (European Environment Agency [EEA], 2018). However, the region experiences significant seasonal water availability variability, exacerbated by climate change. For instance, Central Europe has witnessed increased frequency and intensity of droughts and heatwaves over the past few decades, impacting river flows and groundwater levels (Stahl et al., 2016). In particular, countries like Hungary, Poland, and the Czech Republic are vulnerable to water shortages during prolonged dry periods, which can strain water supplies for agriculture and domestic use.

Climate change projections indicate that the region may experience more pronounced hydrological extremes, including both floods and droughts, leading to increased pressure on existing water management systems (Kundzewicz et al., 2018). The variability in precipitation and rising temperatures contribute to altered snowmelt patterns in mountainous areas like the Carpathians and the Alps, which affect downstream water availability. This variability underscores the need for adaptive water management strategies that can accommodate the changing climatic conditions and reduce the vulnerability of water-dependent sectors.

Climate change poses additional challenges to water security in Eastern and Central Europe. The region is experiencing rising temperatures, altered precipitation patterns, and increased frequency of extreme weather events such as droughts and floods, which impact water availability and quality (Kovats et al., 2014). For example, Central European countries like Poland, Czech Republic, and Hungary have been facing recurrent droughts in recent years, which have reduced river flows and groundwater levels, affecting both agriculture and drinking water supplies (WWF, 2021). On the other hand, extreme flooding, particularly in Eastern European countries like Ukraine and Romania, has led to the contamination of water sources, as floodwaters often carry pollutants and untreated sewage into rivers and reservoirs.

The combination of climate change impacts and existing vulnerabilities in water management systems exacerbates the hardship faced by the population in accessing potable water. The European Commission (2021) has highlighted the need for increased resilience and adaptation strategies, including modernizing water infrastructure, implementing better water management policies, and promoting sustainable agricultural practices to protect water resources.

4.2.2 Water scarcity and infrastructure issues

Access to potable water in Eastern and Central Europe varies significantly between urban and rural areas. In many parts of rural Romania, Moldova, and Ukraine, outdated water infrastructure poses a severe challenge. These regions often rely on wells and local water sources that are prone to contamination, leading to health risks. According to the World Health Organization (WHO, 2020), over 20% of rural populations in Romania and Ukraine lack access to safe drinking water, with many using untreated sources due to a lack of municipal water systems. Infrastructure in these areas is often neglected, with limited investment and maintenance, exacerbating the difficulty of ensuring clean water access for vulnerable communities (OECD, 2021).

In urban areas, access to potable water is generally more reliable, yet challenges remain, particularly in cities with aging infrastructure. For instance, many parts of Bulgaria and Poland face frequent disruptions in water supply due to the deterioration of old water pipes and treatment facilities. These issues not only affect the availability of water but also contribute to contamination risks. The European Environmental Agency (EEA, 2018) notes that over 30% of water infrastructure in Eastern European cities is outdated, leading to losses in water supply and increased costs for households and municipalities.

4.2.3 Pollution and environmental degradation

Pollution is a significant factor contributing to the hardship of accessing potable water in Eastern and Central Europe. Industrial activities, agricultural runoff, and inadequate wastewater treatment facilities are key sources of water contamination. The Danube River, which flows through several countries, including Hungary, Slovakia, and Romania, is one of Europe's most polluted rivers due to industrial waste, pesticides, and agricultural nutrient runoff (ICPDR, 2019). This pollution affects drinking water sources and has broader environmental impacts, threatening aquatic ecosystems and biodiversity.

The legacy of the Soviet era still haunts many parts of Eastern Europe, where heavy industries left behind contaminated sites and polluted water bodies. For example, Moldova and Ukraine have numerous areas where groundwater is contaminated with heavy metals and chemicals due to the lack of proper industrial waste management during the Soviet period (UNDP, 2020). The poor drinking water quality in these areas has led to increased health risks, including gastrointestinal diseases and other waterborne illnesses, particularly affecting low-income and rural populations.

4.2.4 Socio-economic and infrastructural challenges

Eastern and Central Europe face several socio-economic and infrastructural challenges that impact their water security. Despite having relatively abundant water resources, many regions struggle with aging and inefficient water infrastructure. This includes outdated water supply networks, inadequate wastewater treatment facilities, and inefficient irrigation systems, which lead to significant water losses and reduced water quality (World Bank, 2021). For example, in Romania and Bulgaria, high levels of water leakage from distribution systems contribute to water stress, particularly in urban areas with growing populations (European Commission, 2019).

Moreover, the region's transition from centrally planned economies to market-based systems has led to changes in land use and industrial practices, contributing to water pollution from agricultural runoff, mining activities, and industrial discharges. This pollution poses a risk to both surface water and groundwater quality, limiting the availability of safe drinking water (EEA, 2020). Addressing these challenges requires substantial investments in modernizing water infrastructure, improving water efficiency, and implementing stricter pollution control measures.

Socio-economic disparities further complicate access to potable water in Eastern and Central Europe. Rural communities, especially in countries like Bulgaria, Romania, and Moldova, are often marginalized in terms of water service provision due to economic constraints and limited governmental support (Börkey & Lallana, 2018). Additionally, the high cost of upgrading water infrastructure is a significant barrier for many countries in the region, where budget constraints and political instability can limit the prioritization of water-related investments.

Policy and governance issues also contribute to the ongoing challenges in water access. Many countries in Eastern Europe face fragmented water management policies and lack coordination between local and national authorities, which hinders the effective implementation of water protection measures (Varis et al., 2019). Efforts to improve water security require comprehensive strategies that address both the physical infrastructure needs and the socio-political dimensions of water management, including cross-border cooperation for shared water resources like the Danube and Dniester rivers.

The hardship of accessing potable water in Eastern and Central Europe is a complex issue influenced by outdated infrastructure, pollution, climate change impacts, and socio-economic inequalities. Addressing these challenges requires a multi-faceted approach, including investing in modern water infrastructure, strengthening environmental regulations, promoting sustainable water management practices, and enhancing regional cooperation. By tackling these issues, the region can work towards ensuring equitable access to safe drinking water for all its citizens, thus improving public health and socio-economic development.

4.2.5 Governance and policy implications

Fragmented institutional frameworks and transboundary water management issues often complicate the governance of water resources in Eastern and Central Europe. The Danube River Basin, one of the region's most significant shared water resources, involves cooperation among 14 countries, each with its water policies and priorities (International Commission for the Protection of the Danube River [ICPDR], 2022). Effective management of transboundary water resources requires robust cooperation mechanisms, harmonized regulations, and integrated water resource management (IWRM) practices.

Efforts to enhance water security in the region have been supported by the European Union's Water Framework Directive, which aims to achieve good ecological and chemical status for all water bodies by 2027 (European Union, 2000). However, implementation challenges persist, particularly in non-EU Eastern European countries, where regulatory frameworks and enforcement capacities are often weaker (WWF, 2021). Strengthening governance and fostering regional collaboration are essential for addressing water security challenges, especially in transboundary contexts where upstream activities can significantly impact downstream water availability and quality.

4.2.6 Pathways to enhanced water security

To improve water security in Eastern and Central Europe, a multi-faceted approach is needed. This includes investing in modern water infrastructure, promoting sustainable water use practices, and enhancing climate resilience. For instance, increasing the efficiency of irrigation systems, adopting water-saving technologies, and implementing nature-based solutions, such as wetland restoration, can help mitigate the impacts of droughts and floods (EEA, 2019). Furthermore, raising public awareness about water conservation and promoting alternative water sources, such as rainwater harvesting and treated wastewater, can contribute to more sustainable water management.

In conclusion, while Eastern and Central Europe are currently in moderate water security, the region faces significant challenges related to climate variability, infrastructural limitations, and governance complexities.

Addressing these issues will require coordinated efforts at national and regional levels and investments in sustainable water management practices to ensure long-term water security.

4.3 Freshwater Challenges in Eastern and Central Europe: Country-Specific Cases

Eastern and Central Europe encompass diverse hydrological landscapes, where countries face unique challenges related to freshwater management. Despite being relatively water-rich compared to regions like the Middle East or North Africa, these countries encounter pollution, inefficient water use, climate variability, and aging infrastructure, which threaten water security. This section explores country-specific cases in Eastern and Central Europe, highlighting the unique freshwater challenges that Poland, Hungary, Romania, and the Czech Republic face.

4.3.1 Poland: Groundwater contamination and drought

Poland is relatively endowed with water resources but ranks among the lowest in Europe in per capita renewable freshwater availability, largely due to its high population and industrial demands (European Environment Agency [EEA], 2020). One of the significant challenges in Poland is groundwater contamination, exacerbated by intensive agriculture. The overuse of fertilizers and pesticides has led to high levels of nitrates in groundwater, affecting drinking water quality, especially in rural areas (Kowalski et al., 2019). Additionally, Poland faces growing concerns over droughts, particularly in its central and western regions. In recent years, the country has experienced severe summer droughts that have negatively impacted agriculture, leading to significant economic losses (Wójcik & Miętus, 2021). These conditions are expected to worsen due to climate change, prompting the need for improved water management practices, such as sustainable agriculture and drought-resistant crop varieties.

4.3.2 Hungary: Dependence on transboundary water sources

Hungary is highly dependent on transboundary water resources, with over 95% of its surface water coming from neighboring countries through major rivers like the Danube and Tisza (ICPDR, 2022). This dependence makes Hungary particularly vulnerable to upstream activities, including pollution and water extraction, which can impact water quality and availability. One of Hungary's most pressing freshwater challenges is managing water quality in its rivers and lakes, especially given the agricultural runoff and industrial discharges that pollute its waterways (Vári & Ferencz, 2019). Lake Balaton, the largest freshwater lake in Central Europe, has faced issues related to eutrophication due to nutrient inflows from agricultural areas, affecting tourism and local biodiversity (Gyalistras et al., 2020). Hungary's reliance on transboundary water cooperation is thus crucial for its water security, emphasizing the need for regional agreements and collaboration under frameworks like the Danube River Protection Convention.

4.3.3 Romania: Pollution and infrastructure deficiencies

Romania is rich in freshwater resources, particularly through the Danube River and its extensive network of tributaries. However, the country struggles with significant water pollution challenges, primarily due to untreated industrial wastewater, mining activities, and inadequate sewage systems (European Commission, 2019). The Jiu and Argeș rivers, crucial for domestic and industrial water supply, have been reported to suffer from high pollution levels, affecting both human health and aquatic ecosystems (Mihu-Pintilie et al., 2021). Furthermore, Romania faces infrastructural challenges, as much of its water supply and sewage infrastructure is outdated, leading to water losses and inefficiencies (World Bank, 2021). Addressing these issues requires substantial investments in modernizing water infrastructure and enforcing stricter environmental regulations.

4.3.4 Czech Republic: Balancing industrial and environmental water use

The Czech Republic faces the dual challenge of supporting its highly industrialized economy while maintaining environmental water quality. The country is known for its significant industrial water use, particularly in energy production and manufacturing (Czech Statistical Office, 2022). However, this industrial activity often leads to water pollution, especially in the Elbe and Vltava River basins, where contaminants from factories and thermal power plants are prevalent (Matoušková et al., 2019) Additionally, the Czech Republic has experienced frequent droughts over the last decade, which have reduced water levels in rivers and reservoirs, affecting both water supply and hydroelectric power generation (Trnka et al., 2020). The government has responded by implementing measures to increase water retention in the landscape, such as building small water reservoirs and promoting soil conservation techniques to reduce runoff and improve groundwater recharge.

4.4 Addressing Freshwater Challenges: Regional Cooperation and Policy Measures

Eastern and Central European countries are increasingly recognizing the need for integrated water resource management (IWRM) and cross-border cooperation to address their freshwater challenges. The European Union's Water Framework Directive (WFD) has been a critical policy tool in promoting sustainable water use, improving water quality, and enhancing water management across member states (European Union, 2000). However, non-EU countries like Ukraine and Moldova, which are part of the broader Eastern European region, face difficulties aligning their water policies with EU standards, often due to limited financial and technical capacities (WWF, 2021).

To address these challenges, countries must invest in modern water infrastructure, enforce pollution control measures, and promote water-efficient technologies. Regional initiatives, such as the International Commission for the Protection of the Danube River (ICPDR), play a vital role in fostering cooperation among countries sharing water resources, ensuring sustainable use, and protecting aquatic ecosystems (ICPDR, 2022). Additionally, integrating climate adaptation strategies, such as enhancing drought resilience and improving water storage capacity, is crucial for safeguarding water security in future climate uncertainties.

4.5 Water Resources in Eastern and Central Europe: Abundance, Pollution, and Distribution Challenges

Eastern and Central Europe are regions characterized by relatively abundant freshwater resources compared to other parts of the world. This availability is largely due to the extensive network of rivers, lakes, and aquifers that span countries like Poland, Hungary, Romania, and the Czech Republic (European Environment Agency [EEA], 2020). However, despite this natural advantage, the region faces significant challenges related to pollution and the uneven distribution of water resources, undermining water security and access.

4.5.1 Abundant water resources and pollution issues

The region's water resources are extensive, with major rivers such as the Danube, Vistula, and Oder providing critical water supplies for drinking, agriculture, industry, and energy production. However, water pollution remains a persistent issue. Industrial discharges, agricultural runoff, and insufficiently treated urban wastewater contribute to the degradation of water quality (EEA, 2021). For instance, in the Czech Republic, industrial activities have historically led to elevated levels of heavy metals and other contaminants in rivers, necessitating stringent water quality monitoring and rehabilitation efforts (Matoušková et al., 2019). Similarly, Poland's waterways have been affected by high nutrient loads from agricultural fertilizers, leading to eutrophication and the decline of aquatic ecosystems (Nowak et al., 2020).

Romania faces notable pollution in its river systems, particularly the Danube, which is affected by agricultural chemicals, untreated sewage, and industrial waste. The lack of modern wastewater treatment infrastructure in some areas exacerbates these problems, leading to water contamination affecting human health and biodiversity (European Commission, 2019). Moreover, in Hungary, nutrient pollution from agriculture remains a significant concern, affecting both surface and groundwater quality, impacting drinking water supplies (ICPDR, 2022).

4.5.2 Uneven distribution of water resources

While the region enjoys relatively abundant water resources overall, there are pronounced disparities in water availability across different areas. Countries like Slovakia and the Czech Republic generally have good access to fresh water, while others, particularly in southeastern parts of Europe, face seasonal water stress (Trnka et al., 2020). These disparities are largely due to natural geographical variations, but human factors, such as water management policies and infrastructure investment also influence them.

For example, Hungary and Romania experience regional variations in water availability, with some areas suffering from periodic droughts that deplete river levels and aquifers, thus reducing water supply for both urban and agricultural uses (Vári & Ferencz, 2019). In Bulgaria, rural areas often face water shortages due to outdated infrastructure and limited investment in modern water management technologies, leading to inequitable access to water (UNECE, 2020). Additionally, Poland faces challenges in water distribution, where northern and central regions periodically experience droughts, affecting agriculture and increasing competition for water among different sectors (Wójcik & Miętus, 2021).

4.5.3 Strategies for addressing pollution and distribution challenges

To address these challenges, Eastern and Central European countries are increasingly focusing on modernizing their water management infrastructure and implementing stricter environmental regulations. The European Union's Water Framework Directive has been a significant driver in pushing member states to improve water quality through enhanced monitoring, better treatment facilities, and the reduction of pollutant discharges (European Commission, 2021). For instance, the Czech Republic and Slovakia have invested in upgrading wastewater treatment plants to reduce industrial pollution. At the same time, Poland has launched initiatives to protect its groundwater reserves and enhance drought resilience (World Bank, 2021).

Improving water distribution requires infrastructure investments and integrated water management approaches that account for both natural water cycles and human demands. Regional cooperation, particularly in transboundary river basins like the Danube, is crucial for managing shared water resources sustainably and equitably (ICPDR, 2022). Additionally, adopting water-saving technologies and sustainable agricultural practices can help mitigate the impact of pollution and ensure more efficient water use, thus improving the overall water security of the region.

4.6 The Need for Modernized Infrastructure and Addressing Regional Disparities in Water Access in Eastern and Central Europe

Water security in Eastern and Central Europe is increasingly under strain due to outdated infrastructure, socio-economic disparities, and regional inequalities in access to clean water. While this region generally benefits from abundant water resources, inefficient management, aging infrastructure, and uneven distribution of water services highlight the pressing need for modernization and policy reform. This section delves into the implications of these challenges and emphasizes infrastructure modernization's critical role in mitigating regional water access disparities.

4.6.1 Aging infrastructure and its consequences

One of the most significant challenges facing Eastern and Central European countries is the deterioration of water infrastructure, much of which dates back to the mid-20th century. Decades of underinvestment in maintenance and upgrades have resulted in aging pipelines, leaking water supply systems, and inefficient wastewater treatment facilities (European Environment Agency [EEA], 2020). For example, in countries like Poland and Romania, water losses due to leaks in outdated distribution networks can reach up to 30-40%, significantly reducing the availability of potable water for consumers (World Bank, 2021). The failure to modernize infrastructure leads to substantial water losses and increases the risk of contamination, as outdated systems are more prone to leaks that can introduce pollutants into the drinking water supply.

Moreover, old and inefficient wastewater treatment plants often fail to meet European Union standards, resulting in untreated or inadequately treated sewage being discharged into rivers and lakes. This contributes to water pollution, which poses serious environmental and public health risks (European Commission, 2019). In countries like Hungary and Slovakia, outdated sewage systems have led to periodic contamination of water bodies, impacting both rural and urban populations (Vári & Ferencz, 2019). These challenges underscore the need for substantial investments in modernizing water infrastructure to ensure safe and reliable water supply and sanitation services.

4.6.2 Regional disparities in water access

The disparities in water access between urban and rural areas are another critical issue in Eastern and Central Europe. Urban centers generally have better access to modern water supply systems and sanitation facilities, while rural regions, particularly in countries like Romania and Bulgaria lag significantly behind (United Nations Economic Commission for Europe [UNECE], 2020). For instance, in Romania, nearly 30% of the rural population still relies on private wells, many of which are contaminated with nitrates and bacteria due to inadequate protection and regulation (Mihu-Pintilie et al., 2021). This disparity is further exacerbated by socio-economic factors, with poorer regions facing greater challenges in accessing safe drinking water and sanitation services.

Additionally, in the Czech Republic and Slovakia, small towns and rural communities often struggle with limited financial resources to invest in water infrastructure, leading to inadequate service provision and higher vulnerability to water-related challenges (Czech Statistical Office, 2022). The uneven distribution of water resources and infrastructure not only affects public health but also limits economic development opportunities in disadvantaged regions, perpetuating a cycle of poverty and underdevelopment.

4.6.3 Climate change and the increasing need for resilient infrastructure

Climate change is expected to exacerbate water-related challenges in Eastern and Central Europe, with increasing frequency of droughts, floods, and temperature extremes. These climatic changes place additional stress on already overburdened water infrastructure (Trnka et al., 2020). For example, Hungary has experienced recurring droughts that have reduced water levels in rivers and reservoirs, impacting both drinking water supplies and agricultural productivity (ICPDR, 2022). The lack of resilient infrastructure to store and manage water during dry periods highlights the urgent need for investments in modern water management solutions, such as rainwater harvesting systems, drought-resistant water distribution networks, and enhanced groundwater recharge techniques.

In response to these challenges, several countries have started implementing measures to enhance their water management capabilities. Poland, for example, has launched initiatives to modernize its water supply networks and build new reservoirs to improve water retention during periods of drought (Wójcik & Mi tus, 2021). Similarly, the Czech Republic has been investing in improving water treatment technologies to reduce industrial pollution in its rivers, ensuring a cleaner and more reliable water supply for its citizens (Matoušková et al., 2019).

4.6.4 Policy recommendations for closing the infrastructure gap

Addressing the infrastructure and regional disparities in water access requires a multi-faceted approach that includes national and regional strategies. First, substantial investments in modernizing water infrastructure are essential. This includes replacing old pipelines, upgrading wastewater treatment plants, and expanding access to safe drinking water in rural and underserved areas (World Bank, 2021). The European Union's Cohesion Fund and the Green Deal offer financial support to member states for improving water infrastructure, which can be instrumental in bridging the investment gap (European Union, 2020).

Second, enhancing regional cooperation on water management, particularly in transboundary river basins like the Danube, can promote shared solutions to common challenges. Initiatives like the Danube River Protection Convention provide a platform for countries to collaborate on sustainable water management practices and pollution control (ICPDR, 2022). Strengthening these cooperative frameworks can help countries address both current and future water security challenges more effectively.

Lastly, implementing policies that prioritize climate adaptation and water conservation is crucial. Governments should focus on promoting water-efficient technologies, sustainable agricultural practices, and nature-based solutions such as wetland restoration to enhance water retention and reduce the impact of extreme weather events (WWF, 2021).

4.7 Southeast Asia

Southeast Asia, a subregion of Asia, is situated between the Indian and Pacific Oceans and comprises diverse landscapes, cultures, and economies. This region includes mainland countries like Myanmar, Thailand, Laos, Cambodia, and Vietnam and archipelagic nations such as Indonesia, the Philippines, Malaysia, Singapore, Brunei, and Timor-Leste. Despite being endowed with significant water resources, Southeast Asia faces considerable challenges in providing potable water to its growing population. The total population of Southeast Asia is over 680 million, with many residing in rural and underserved urban areas where water supply infrastructure is inadequate (UNESCAP, 2021). Access to clean water is essential for health, livelihoods, and socio-economic development. Yet, millions across the region struggle to secure safe drinking water due to a combination of environmental, infrastructural, and socio-economic factors.

4.7.1 Water security in Southeast Asia: Abundance and challenges

Southeast Asia is a region that enjoys significant water resources due to its tropical climate and abundant rainfall, with major river systems like the Mekong, Chao Phraya, and Irrawaddy providing vital water supplies for agriculture, industry, and domestic use. Despite this relative abundance, many parts of Southeast Asia are classified as having moderate water security due to a combination of factors such as uneven distribution of resources, pollution, infrastructure deficits, and the impacts of climate change (Asian Development Bank [ADB], 2016).

4.7.2 Abundant but unevenly distributed water resources

Southeast Asia's overall water availability is generally high, yet there is a stark disparity in the distribution of these resources. For example, while countries like Laos and Myanmar are water-rich with extensive river networks, others like Thailand and Vietnam face periodic water stress, particularly during the dry season (Molle et al., 2021). The Mekong River, which flows through multiple Southeast Asian countries, is crucial for livelihoods and food security, especially in Cambodia and Vietnam. However, the benefits of this river are not equally shared, with upstream and downstream countries facing different challenges related to water availability and access (Mekong River Commission, 2020).

The uneven distribution of water resources is exacerbated by seasonal variations in rainfall, where the monsoon season brings excessive water, often resulting in flooding, while the dry season can lead to water shortages. Thailand, for instance, experiences significant regional disparities, with the northeastern part of the country being particularly vulnerable to droughts, impacting agricultural productivity and water supply (ADB, 2019). Similarly, Indonesia faces challenges in managing its water resources across its vast archipelago, where some islands suffer from water scarcity while others have an abundance (World Bank, 2021).

Urban areas are not immune to water challenges either. Rapid urbanization has led to the proliferation of informal settlements or slums in cities like Jakarta, Manila, and Bangkok, where access to piped water is limited, forcing residents to rely on alternative, often unsafe, sources. As of 2021, approximately 11 million people in Jakarta alone depend on groundwater extraction, which is frequently contaminated due to inadequate waste management and sanitation (PwC Indonesia, 2021). The lack of reliable water infrastructure and the over-reliance on groundwater are contributing to both water scarcity and quality degradation, particularly in densely populated urban areas.

4.7.3 Pollution and infrastructure challenges

Water pollution is a pressing concern in Southeast Asia, particularly in rapidly urbanizing and industrializing areas. The rapid economic growth of countries like Vietnam, Thailand, and Indonesia has led to increased water contamination from industrial effluents, agricultural runoff, and untreated urban wastewater (Hoang & Yabe, 2019). In Vietnam, the Red River and Mekong Delta regions have seen significant water quality deterioration due to pesticide use and industrial discharges, affecting both drinking water and agricultural irrigation (Nguyen et al., 2018). Similarly, in Malaysia river pollution from palm oil processing and other industries poses a threat to water security, impacting communities and ecosystems (Aziz et al., 2020).

The lack of adequate infrastructure compounds these issues. Many countries in the region need more water treatment facilities, which limits their ability to provide safe and reliable water to urban and rural populations (ADB, 2016). For instance, Indonesia's water supply infrastructure faces significant challenges due to aging pipelines and inadequate water treatment capacity, leading to issues in both water quality and access, particularly in rural and remote areas (World Bank, 2021).

4.7.4 Climate change and its impacts on water security

Environmental factors further exacerbate the potable water crisis in Southeast Asia. The region is highly vulnerable to the impacts of climate change, including rising temperatures, changing precipitation patterns, and increased frequency of extreme weather events such as typhoons and droughts (Hijioka et al., 2014). These changes have severe implications for water availability and quality. For instance, Vietnam's Mekong Delta, a critical agricultural and freshwater source, is experiencing saltwater intrusion due to rising sea levels, which threatens both drinking water supplies and food security (UNESCO, 2022). Similarly, Thailand and the Philippines face recurrent droughts that reduce freshwater availability, impacting millions of rural and urban households reliant on surface and groundwater sources.

Deforestation, driven by agricultural expansion and logging, has also disrupted natural water cycles, reducing the availability of freshwater resources (Koh & Ghazoul, 2010). Forest loss diminishes the land's ability to absorb and filter water, increasing runoff and reducing groundwater recharge, which is critical for maintaining water supplies during dry seasons. The combination of deforestation and climate change has placed significant pressure on water systems across the region, complicating efforts to provide safe drinking water.

Climate change presents an additional layer of complexity to water security in Southeast Asia. The region is highly vulnerable to the impacts of climate change, which include more frequent and severe droughts, floods, and sea-level rise (IPCC, 2021). These changes exacerbate the existing challenges of water management, particularly in low-lying areas like the Mekong

Delta, which is already experiencing salinization of freshwater resources due to rising sea levels (Hoang et al., 2020).

In Thailand and the Philippines, increasingly erratic weather patterns have resulted in both droughts and floods that threaten water security, agriculture, and livelihoods (Rosales & Sy, 2020). The Philippines, which is frequently hit by typhoons, faces severe disruptions to its water supply systems due to extreme weather events, highlighting the need for climate-resilient water management strategies (Lagmay et al., 2019).

4.7.5 Socio-economic inequities and policy challenges

Socio-economic disparities play a significant role in the hardship of accessing potable water. While wealthier urban residents may have access to private water supplies or bottled water, poorer communities often rely on untreated water sources, leading to higher incidences of waterborne diseases (Jha, 2019). The World Health Organization (WHO, 2020) notes that inadequate access to clean water, sanitation, and hygiene is a major public health issue in Southeast Asia, contributing to the spread of diseases such as cholera, typhoid, and diarrhea. The economic costs of these water-related health problems are substantial, further entrenching poverty and inequality in affected communities.

Policy and governance challenges also hinder efforts to improve water access. Fragmented regulatory frameworks, inadequate investment in water infrastructure, and corruption are significant barriers to ensuring equitable water distribution (Tortajada, 2016). Effective water management in Southeast Asia requires a coordinated approach that includes the modernization of infrastructure, enforcement of environmental regulations, and investment in sustainable water supply solutions. Collaborative efforts between governments, non-governmental organizations, and the private sector are essential for addressing the complex and multifaceted water challenges facing the region.

The potable water crisis in Southeast Asia is a multifaceted issue rooted in geographical, environmental, and socio-economic factors. Despite the region's abundant water resources, millions of people need help accessing safe drinking water due to infrastructure limitations, pollution, climate change impacts, and socio-economic disparities. Addressing these challenges requires an integrated approach that combines technological solutions, sustainable resource management, and policy reforms to ensure that all urban and rural communities have access to safe and reliable water supplies. As Southeast Asia continues to grow and develop, prioritizing water security will be critical to achieving sustainable development and improving the quality of life for its diverse populations.

4.7.6 Strategies for improving water security

Addressing water security in Southeast Asia requires a multifaceted approach that includes infrastructure modernization, improved water management policies, and regional cooperation. The Asian Development Bank has emphasized the need for investment in water infrastructure, such as dams, reservoirs, and irrigation systems, to enhance water storage and distribution, particularly in water-stressed areas (ADB, 2016). Additionally, there is a growing recognition of the need for integrated water resources management (IWRM) to ensure the sustainable use of water resources across different sectors (Molle et al., 2021).

Countries like Singapore have made significant strides in improving water security through innovative solutions such as water recycling, desalination, and rainwater harvesting, setting an example for other nations in the region (Tortajada & Joshi, 2013). Regional cooperation, particularly in transboundary river basins like the Mekong, is crucial for sustainable water management, as it helps mitigate conflicts over shared resources and promotes collective action in addressing pollution and climate impacts (Mekong River Commission, 2020).

4.8 Comparative Analysis of Water Security in Southeast Asia

Southeast Asia, known for its lush landscapes and abundant water resources, paradoxically faces significant water security challenges. This region includes countries with diverse geographical and climatic conditions, such as Indonesia, Malaysia, Thailand, Vietnam, the Philippines, Myanmar, Cambodia, Laos, Singapore, Brunei, and Timor-Leste. The level of water security varies widely across these nations due to differences in water resource availability, mana gement practices, socio-economic development, and the impacts of climate change (Asian Development Bank [ADB], 2016). Despite being water-abundant, Southeast Asia faces a paradox where rapid population growth, industrialization, pollution, and climate change stress water resources, resulting in varied water security profiles across countries.

4.8.1 Indonesia: high water availability with regional disparities

Indonesia, as the largest archipelago in the world, has abundant water resources, with an average annual rainfall of 2,700 mm. However, water distribution is uneven, with some areas experiencing water shortages, particularly during the dry season. Java, the most populous island, faces water stress due to high demand from agriculture, industry, and households (World Bank, 2021). The combination of rapid urbanization and poor water infrastructure has led to challenges in water quality, with pollution from untreated wastewater and industrial discharges affecting rivers and groundwater (Amarasinghe & Smakhtin, 2014). The country's water security is further threatened by climate change, which exacerbates floods and droughts.

4.8.2 Vietnam: Mekong delta under threat

Vietnam relies heavily on the Mekong River, especially for agriculture and fisheries. However, water security is threatened due to upstream dam construction, which alters flow regimes, and saltwater intrusion in the Mekong Delta, driven by rising sea levels (Mekong River Commission, 2020). The Red River Basin in northern Vietnam also faces water quality issues due to industrial pollution and agricultural runoff, affecting drinking water and food production (Nguyen et al., 2018). The Vietnamese government has been improving water infrastructure and adopting integrated water resource management (IWRM) strategies to cope with these challenges (ADB, 2019).

4.8.3 Thailand: Droughts and floods

Thailand's water security is characterized by seasonal variability, with severe floods during the rainy season and droughts during the dry season. The northeastern region, known as Isaan, is particularly prone to droughts, impacting agriculture and livelihoods (Molle & Floch, 2021). Additionally, pollution from agricultural runoff, particularly from intensive rice and sugarcane cultivation, has led to deteriorating water quality in the Chao Phraya River Basin (Rigg et al., 2019). The Thai government has invested in large-scale irrigation projects and reservoirs to mitigate water scarcity, but challenges remain in managing these resources efficiently.

4.8.4 Malaysia: Water-rich but facing pollution

Malaysia enjoys abundant freshwater resources, but water security issues arise primarily from pollution and urbanization. River pollution due to palm oil processing, industrial waste, and urban runoff has impacted water quality and biodiversity (Aziz et al., 2020). The Klang Valley, which includes Kuala Lumpur, faces significant water stress due to high demand and pollution. The Malaysian government has been actively addressing these issues by investing in water treatment facilities and promoting stricter environmental regulations (Rahman et al., 2018).

4.8.5 The Philippines: Vulnerability to natural disasters

The Philippines is particularly vulnerable to climate change impacts such as typhoons and heavy rainfall, often leading to flooding and water supply disruptions (Lagmay et al., 2019). Manila, the capital, faces water shortages due to inadequate infrastructure and pollution of major water bodies like Laguna de Bay and the Pasig River (ADB, 2016). Rural areas also suffer from limited access to clean water, exacerbating social inequalities. The government has initiated several water management projects to improve water access and resilience, including rainwater harvesting and desalination (World Bank, 2021).

4.8.6 Singapore: A model of water security

Despite having limited natural freshwater resources, Singapore has achieved high water security through innovative approaches such as water recycling (NEWater), desalination, and rainwater harvesting (Tortajada & Joshi, 2013). The country's commitment to water self-sufficiency has made it a global model in water management, demonstrating the effectiveness of technology and policy in addressing water scarcity (Public Utilities Board, 2020). Singapore's approach focuses on the "Four National Taps" strategy, which ensures diverse sources of water to meet the needs of its population.

4.8.7 Cambodia and Laos: Abundant water but limited infrastructure

Cambodia and Laos, while having abundant water resources, particularly from the Mekong River, need more infrastructure and stronger regulatory frameworks. These countries face challenges in providing safe drinking water and sanitation, particularly in rural areas (ADB, 2016). The reliance on the Mekong River for agriculture and fisheries makes them vulnerable to upstream developments and climate change impacts, such as altered rainfall patterns and rising temperatures (Mekong River Commission, 2020).

4.8.8 Myanmar: Untapped potential amidst challenges

Myanmar has significant water resources, but political instability and underinvestment in water infrastructure have hindered its ability to manage these resources effectively (ADB, 2019). The Irrawaddy River is the lifeline for many communities, yet it faces pollution from mining activities and sedimentation due to deforestation (Myint et al., 2020). Addressing these challenges requires improved governance, investment in infrastructure, and sustainable water management practices.

A complex interplay of natural abundance, human activities, and socio-political factors shapes Southeast Asia's water security landscape. While some countries, like Singapore, have excelled in achieving water security through innovative management strategies, others face significant challenges due to pollution, infrastructure deficits, and climate change. Addressing these disparities requires regional cooperation, infrastructure investment, and sustainable water management practices to ensure water security for all (Hoanh et al., 2020).

4.9 Generally Good Rainfall but Periodic Water Stress Due to Monsoon Variability and Pollution

Southeast Asia is characterized by abundant freshwater resources, supported by heavy rainfall averaging between 1,500 to 3,000 mm per year (ADB, 2016). This rainfall, combined with major river systems like the Mekong, Irrawaddy, Chao Phraya, and Red Rivers, provides the region with considerable water resources. However, despite this abundance, many Southeast Asian countries face periodic water stress due to a combination of monsoon variability, uneven rainfall distribution, pollution, and growing water demand.

4.9.1 Monsoon variability and periodic water stress

One of the key challenges in Southeast Asia is the high dependence on monsoon rains, which can be unpredictable and vary greatly from year to year. The region experiences a tropical monsoon climate, resulting in distinct wet and dry seasons (Hoanh et al., 2020). In some years, delayed or insufficient monsoon rains can lead to drought, particularly in Thailand, Vietnam, and Indonesia. For instance, the El Niño phenomenon has historically led to reduced rainfall in the region, impacting water availability for agriculture, which is a critical sector for the economies of these nations (Mekong River Commission, 2020).

In countries like Thailand, the northeast region (Isaan) is especially vulnerable to water shortages during the dry season, affecting both drinking water supply and agricultural production (Molle & Floch, 2021). Vietnam's Mekong Delta, known as the country's " rice bowl, " faces similar challenges during drought years, with reduced water flow impacting rice paddies and aquaculture (Nguyen & Bach, 2018). Meanwhile, Indonesia's Java island, home to over half of the nation's population, experiences periodic droughts that strain its water resources due to high agricultural and urban demand (World Bank, 2021).

4.9.2 Pollution and water quality degradation

In addition to issues related to water availability, Southeast Asia faces significant water quality challenges. Rapid urbanization, industrialization, and agricultural expansion have led to increased pollution of water bodies across the region. Major rivers, such as the Chao Phraya in Thailand and the Pasig River in the Philippines, suffer from high levels of pollution due to untreated sewage, industrial waste, and agricultural runoff (Rigg et al., 2019). These pollutants degrade water quality and threaten aquatic ecosystems and public health.

Countries like Malaysia and the Philippines experience severe river pollution, especially in densely populated urban centers. For instance, Malaysia's Klang River has been significantly impacted by urban runoff and industrial waste, compromising its use for drinking water and other purposes (Aziz et al., 2020). Similarly, the Philippines faces water quality issues in Manila due to heavy pollution in the Laguna de Bay and Pasig River, leading to challenges in providing safe drinking water (ADB, 2016).

4.9.3 Uneven distribution of water resources

Despite the overall abundance of water resources, there is significant spatial and temporal variability in water availability across the region. This uneven distribution exacerbates water security challenges, particularly in countries with diverse topographies and climates. For example, while Indonesia and Malaysia have relatively abundant water resources, these are unevenly distributed across their islands. Java, Indonesia's most populated island, often faces water shortages, while outer islands like Kalimantan and Sumatra have surplus water that is not fully utilized (Amarasinghe & Smakhtin, 2014).

Similarly, in the case of Vietnam, the northern and southern regions are more water-rich compared to the central regions, which are prone to water shortages during the dry season (ADB, 2019). Cambodia and Laos, while having considerable water resources due to the Mekong River, struggle with infrastructure challenges, limiting their ability to harness these resources effectively, especially during the dry season when water levels drop (Mekong River Commission, 2020).

Southeast Asia's water security is marked by a paradox of abundance and scarcity. The region generally enjoys high levels of rainfall but faces periodic water stress due to monsoon variability, pollution, and uneven distribution of water resources. Addressing these challenges requires a multi-faceted approach, including investments in water infrastructure, pollution control, and adopting integrated water resource management (IWRM) practices to ensure sustainable water use for future generations (Hoang & Kummu, 2020). Regional cooperation, especially among countries sharing transboundary rivers like the Mekong, is also crucial for effective water management and mitigating the impacts of climate variability.

4.10 Impact of Water Security Challenges in Southeast Asia: Seasonal Scarcity, Flood Risks, and Wastewater Management

While blessed with abundant rainfall and rich water resources, Southeast Asia faces significant water security challenges due to seasonal variability, flood risks during the monsoon season, and the growing problem of wastewater management. Rapid urbanization, industrialization, and population growth exacerbate these issues, putting increasing pressure on already strained water systems.

4.10.1 Seasonal water scarcity and monsoon dependence

Southeast Asia experiences a tropical monsoon climate characterized by distinct wet and dry seasons. During the wet season, the region receives a significant amount of its annual rainfall, which is critical for replenishing rivers, lakes, and groundwater reserves. However, this reliance on seasonal monsoons leads to challenges in water management, particularly in the dry months when water scarcity becomes a pressing issue (ADB, 2016). Countries like Thailand, Indonesia, and the Philippines often face drought conditions during El Niño years, which can significantly reduce water availability for agriculture, drinking water, and hydroelectric power generation (Hoang & Kummu, 2020).

In Vietnam, the Mekong Delta, a crucial area for rice production, experiences water shortages during the dry season, especially when upstream countries divert water for their own agricultural and energy needs (Mekong River Commission, 2020). Similarly, Indonesia's densely populated Java Island frequently faces dry-season water shortages, impacting urban and agricultural regions (World Bank, 2021). This seasonal scarcity underscores the need for better water storage, conservation, and distribution systems to mitigate the impacts of water stress.

4.10.2 Risk from floods during the monsoon season

The same monsoon rains that alleviate water scarcity can also bring devastating floods, particularly in low-lying and coastal areas. Countries like Thailand, Myanmar, and the Philippines are especially vulnerable to monsoon-induced floods, which can result in significant socio-economic impacts, including loss of life, property damage, and disruption of livelihoods (Rigg et al., 2019). For example, the Chao Phraya River in Thailand and the Cagayan River in the Philippines are prone to overflowing during heavy monsoon rains, leading to widespread flooding in surrounding communities (ADB, 2019).

Flooding results from heavy rainfall, poor urban planning, and deforestation, which reduce the land's ability to absorb water (Molle & Floch, 2021). In urban centers like Jakarta and Bangkok, inadequate drainage systems and rapid urban expansion into flood-prone areas have increased the frequency and severity of urban flooding. Additionally, sea-level rise due to climate change poses a long-term threat to coastal cities in Southeast Asia, further compounding flood risks (Nguyen & Bach, 2018).

4.10.3 Challenges of managing wastewater

As Southeast Asia's population grows and industrializes, managing wastewater has become an increasingly pressing issue. Many countries in the region lack adequate infrastructure for treating domestic and industrial wastewater, leading to severe water pollution. Rivers like the Chao Phraya in Thailand and the Pasig in the Philippines are heavily polluted with untreated sewage, industrial discharges, and agricultural runoff, compromising water quality and public health (Aziz et al., 2020).

The problem of wastewater management is particularly acute in urban areas where infrastructure development has not kept pace with population growth. For instance, less than half of the wastewater generated in Southeast Asia is treated before being discharged into the environment, resulting in contaminated waterways that can no longer safely support human or aquatic life (Hoanh et al., 2020). The lack of effective wastewater management poses a threat not only to water quality but also to the region's ability to meet the Sustainable Development Goals (SDGs), particularly those related to clean water and sanitation (World Bank, 2021).

Indonesia and Vietnam, for example, have made strides in expanding access to clean water but still face significant challenges in developing comprehensive wastewater treatment systems, especially in rural and peri-urban areas (Amarasinghe & Smakhtin, 2014). Addressing this issue requires substantial investments in infrastructure, policy reforms, and public awareness campaigns to promote water conservation and pollution reduction.

In conclusion, while Southeast Asia enjoys abundant water resources, the region faces critical challenges related to seasonal water scarcity, flood risks during the monsoon season, and inadequate wastewater management. The seasonal variability of water availability highlights the need for better water storage and conservation strategies, while the threat of floods calls for improved infrastructure and urban planning to enhance resilience. Moreover, addressing the growing water pollution problem through improved wastewater treatment is essential for safeguarding the region's water resources and ensuring long-term sustainability. By adopting integrated water resource management (IWRM) approaches, Southeast Asian countries can better navigate the complexities of water security and enhance their resilience to natural and human-made water challenges.

References

Amarasinghe, U. A., & Smakhtin, V. (2014). Water scarcity and future challenges in Southeast Asia. *Water Resources Management*, *28*(6), 1747–1760.

Aqueduct Water Risk Atlas. (2023). *Water risk and scarcity assessment.* World Resources Institute. Retrieved from https://www.wri.org

Asian Development Bank (ADB). (2022). *Asian Water Development Outlook 2022: Advancing Water Security Across Asia and the Pacific*. Manila: Asian Development Bank.

Asian Development Bank. (2016). *Asian Water Development Outlook 2016: Strengthening Water Security in Asia and the Pacific*. Asian Development Bank.

Asian Development Bank. (2019). *Flood Risk Management and Mitigation in Southeast Asia*. Asian Development Bank.

Asian Development Bank. (2019). *Water Security in Thailand: Addressing Challenges of Drought and Floods*. Asian Development Bank.

Aziz, N. A. A., Abdullah, S., & Mohammad, F. (2020). Palm oil industry and its impact on river pollution in Malaysia. *Environmental Science and Pollution Research*, *27*(16), 19875–19889.

Aziz, N. A. A., Abdullah, S., & Mohammad, F. (2020). River pollution from palm oil industry in Malaysia: Environmental impacts and control strategies. *Environmental Science and Pollution Research*, *27*(15), 19860–19874.

Börkey, P., & Lallana, C. (2018). *Improving the Governance of Water Management in Eastern Europe, the Caucasus and Central Asia*. Paris: OECD Publishing.

Czech Statistical Office. (2022). *Water management in the Czech Republic: Statistical Yearbook 2022*. Czech Statistical Office.

European Commission. (2019). *Environmental Implementation Review 2019: Country Report - Romania*. European Commission.

European Commission. (2021). *Adapting to Climate Change in Eastern and Central Europe*. Brussels: European Commission.

European Environment Agency. (2018). *European waters: Assessment of status and pressures 2018*. EEA Report No 7/2018.

European Environment Agency. (2019). *Adaptation challenges and opportunities for the European water sector*. EEA Report No 3/2019.

European Environment Agency. (2020). *European waters: Assessment of status and pressures 2020*. EEA Report No 1/2020.

European Environment Agency. (2020). *State of Europe's environment - Water quality*. EEA Report No 11/2020.

European Environment Agency. (2021). *Water quality in Europe: Assessment of rivers, lakes, and groundwater*. EEA Report.

European Environmental Agency (EEA). (2018). *Water Resources in Europe: Facing the Challenge*. Luxembourg: Publications Office of the European Union.

European Union. (2000). *Directive 2000/60/EC of the European Parliament and of the Council establishing a framework for Community action in the field of water policy*. Luxembourg: Publications Office of the European Union.

European Union. (2020). *The European Green Deal*. European Commission.

Falkenmark, M., & Rockström, J. (2022). *Balancing Water for Humans and Nature: The New Approach in Ecohydrology*. Earthscan.

Gyalistras, D., Schmid, M., & Schär, C. (2020). A review of the impacts of climate change on Lake Balaton and possible adaptation strategies. *Aquatic Sciences*, *82*(2), 45–58.

Hijioka, Y., Lin, E., Pereira, J. J., Corlett, R. T., Cui, X., Insarov, G., ... & Surjan, A. (2014). *Asia*. In V. R. Barros, C. B. Field, D. J. Dokken, M. D. Mastrandrea, K. J. Mach, & T. E. Bilir (Eds.), *Climate Change*

2014: Impacts, Adaptation, and Vulnerability. Part B: Regional Aspects. Cambridge University Press.

Hoang, L. P., & Kummu, M. (2020). Water resources management in the Mekong River Basin: Challenges and perspectives. *Water Policy*, *22*(6), 1120–1135.

Hoang, L. P., & Yabe, M. (2019). Water pollution and its impact on agriculture: Evidence from the Red River and Mekong Delta regions. *Water Resources Management*, *33*(4), 1375–1388.

Hoang, L. P., Kummu, M., Lauri, H., et al. (2020). Future salinity intrusion in the Mekong Delta due to sea level rise and river flow changes. *Science of the Total Environment*, *703*, 134580.

Hoanh, C. T., Smakhtin, V., & Johnston, R. (2020). Climate variability and impacts on water resources in Southeast Asia. *Water International*, *45*(3), 215–231.

ICPDR (International Commission for the Protection of the Danube River). (2019). *Danube River Basin Management Plan*. Vienna: ICPDR.

ICPDR. (2022). *Annual Report 2021: Protecting the Danube River Basin*. International Commission for the Protection of the Danube River.

Intergovernmental Panel on Climate Change. (2021). *Climate Change 2021: Impacts, Adaptation, and Vulnerability*. Cambridge University Press.

International Commission for the Protection of the Danube River (ICPDR). (2022). *Annual Report 2021*.

Jha, P. K. (2019). *Water and health in Southeast Asia: Addressing the challenge of equitable access*. World Health Organization.

Koh, L. P., & Ghazoul, J. (2010). *Spatially explicit scenario analysis for reconciling agricultural expansion, forest protection, and carbon conservation in Indonesia*. Proceedings of the National Academy of Sciences, 107(24), 11140–11144.

Kovats, R. S., Valentini, R., Bouwer, L. M., Georgopoulou, E., Jacob, D., Martin, E., ... & Soussana, J. F. (2014). *Europe*. In C. B. Field, V. R. Barros, D. J. Dokken, K. J. Mach, & M. D. Mastrandrea (Eds.), *Climate Change 2014: Impacts, Adaptation, and Vulnerability*. Part B: Regional Aspects. Cambridge University Press.

Kowalski, Z., Makara, A., & Sobczyk, W. (2019). Groundwater pollution and its prevention in Poland. *Environmental Science and Pollution Research*, *26*(7), 6143–6155.

Kummu, M., Ward, P. J., de Moel, H., & Varis, O. (2016). Is physical water scarcity a new phenomenon? Global assessment of water shortage over the last two millennia. *Environmental Research Letters*, *5*(3), 034006.

Kundzewicz, Z. W., Mata, L. J., Arnell, N. W., Döll, P., Jimenez, B., Miller, K., ... & Shiklomanov, I. (2018). Freshwater resources and their management. In M. Parry, O. Canziani, J. Palutikof, & P. van der Linden (Eds.), *Climate Change 2022: Impacts, Adaptation, and Vulnerability* (pp. 173–210). Cambridge University Press.

Lagmay, A. M. F., Aracan, K. A., & Dimalanta, C. B. (2019). Disaster risk reduction strategies for water management in the Philippines. *International Journal of Disaster Risk Reduction, 41*, 101322.

Matoušková, L., et al. (2019). *Heavy metal pollution in the Czech Republic's river systems: Monitoring and mitigation. Journal of Environmental Management, 234*, 101–112.

Matoušková, Z., Hlaváčková, P., & Pivokonský, M. (2019). Industrial water pollution in the Czech Republic and its control measures. *Journal of Environmental Management, 241*, 372–380.

Mekong River Commission. (2020). *State of the Basin Report 2020*. Mekong River Commission.

Mihu-Pintilie, A., Andrei, M., & Cărăuşu, G. (2021). Assessing water quality in Romania's major river basins. *Water Research, 189*, 116620.

Molle, F., & Floch, P. (2021). Water resources in Thailand: Challenges of drought and flood management. *Water Resources Management, 35*(4), 1320–1335.

Molle, F., Floch, P., Promphakping, B., & Blake, D. J. H. (2021). Water use in the Greater Mekong Subregion: Issues and perspectives. *Water Policy, 23*(3), 567–583.

Nguyen, T. P., & Bach, T. S. (2018). Industrial pollution and water quality in Vietnam. *Environmental Monitoring and Assessment, 190*(4), 275.

Nguyen, T. P., & Bach, T. S. (2018). The impact of industrial pollution on water resources in Vietnam. *Environmental Monitoring and Assessment, 190*(5), 291.

Nowak, B., & Wojciechowska, M. (2020). Nutrient loads from agriculture and their impact on water quality in Poland. *Water Resources Management, 34*(8), 2431–2443.

OECD. (2017). *Water Risk Hotspots for Agriculture*. OECD Publishing.

OECD. (2021). *Water Governance in Eastern Europe, Caucasus and Central Asia*. Paris: OECD Publishing.

PwC Indonesia. (2021). *Jakarta's Groundwater Management: Addressing a Sinking Crisis*. Jakarta: PwC.

Rigg, J., Oven, K., & Berger, A. (2019). Urban water insecurity and health in Southeast Asia. *Social Science & Medicine, 240*, 112570.

Rockström, J., Steffen, W., Noone, K., Persson, Å., Chapin, F. S., Lambin, E. F., ... & Foley, J. A. (2014). Planetary boundaries: Exploring the safe operating space for humanity. *Ecology and Society*, *14*(2), 32.

Rosales, J., & Sy, V. A. (2020). Climate variability and water resources in Southeast Asia. *Climatic Change*, *162*(4), 1847–1862.

Stahl, K., Tallaksen, L. M., Hannaford, J., & Van Lanen, H. A. J. (2016). Filling the white space on maps of European runoff trends: Estimates from a multi-model ensemble. *Hydrology and Earth System Sciences*, *20*(6), 2037–2052.

Tortajada, C. (2016). *Water management in Southeast Asia: Issues and perspectives*. International Journal of Water Resources Development, 32(5), 768–776.

Tortajada, C., & Joshi, Y. K. (2013). Water management in Singapore: Achievements and lessons. *Water Resources Development*, *29*(1), 55–71.

Tortajada, C., & Joshi, Y. K. (2013). Water management strategies in Singapore: Innovations and outcomes. *Water Resources Development*, *29*(2), 125–138.

Trnka, M., Balek, J., & Semerádová, D. (2020). Drought impacts and adaptation strategies in the Czech Republic: A case study of recent trends and mitigation measures. *Climate Policy*, *20*(2), 245–262.

Trnka, M., et al. (2020). Drought adaptation strategies in Central Europe: Trends and challenges. *Environmental Research Letters*, *15*(4), 044018.

UNDP (United Nations Development Programme). (2020). *Addressing the Water Crisis in Moldova*. Chisinau: UNDP Moldova.

UNECE. (2020). *Progress on Transboundary Water Cooperation under the Water Convention*. UNECE.

UNESCO. (2022). *The Mekong River and climate change: Impacts on water resources and agriculture*. Paris: UNESCO.

United Nations Economic and Social Commission for Asia and the Pacific (UNESCAP). (2021). *Water Security in Asia-Pacific: Issues and Options*. Bangkok: UNESCAP.

United Nations Economic Commission for Europe. (2020). *Progress on Transboundary Water Cooperation under the Water Convention*. UNECE.

UN-Water. (2023). *What is water security.* United Nations. Retrieved from https://www.unwater.org

Vári, A., & Ferencz, Z. (2019). Addressing water management challenges in Hungary. *Water Policy*, *21*(2), 345–361.

Vári, A., & Ferencz, Z. (2019). Water management challenges in Hungary: The role of policy and practice. *Environmental Policy and Governance*, *29*(6), 433–445.

Varis, O., Kummu, M., & Salmivaara, A. (2019). *Water resources development and management in Eastern Europe*. Journal of Hydrology, 573, 763–775.

Vörösmarty, C. J., Green, P., Salisbury, J., & Lammers, R. B. (2010). Global water resources: Vulnerability from climate change and population growth. *Science*, *289*(5477), 284–288.

Wójcik, R., & Miętus, M. (2021). Droughts in Poland: Frequency, impact, and adaptation measures. *Climate Research*, *86*(1), 47–62.

World Bank. (2020). *Data on access to improved water sources in Southeast Asia*. Washington, D.C.: World Bank.

World Bank. (2021). *Enhancing water security in Poland: Strategies and investments*. World Bank Publications.

World Bank. (2021). *Indonesia: Water Supply and Sanitation Sector Assessment Strategy*. World Bank Publications.

World Bank. (2021). *Modernizing Water Infrastructure in Romania: Enhancing Water Security*. World Bank Publications.

World Bank. (2021). *Water Security and Resilience in Europe and Central Asia: A Call for Action*. World Bank Publications.

World Bank. (2022). *Brazil Water Security Report*. World Bank Publications.

World Health Organization (WHO). (2020). *Drinking Water Quality in Eastern Europe and Central Asia*. Geneva: WHO.

World Health Organization (WHO). (2020). *Guidelines on drinking water quality*. Geneva: WHO

WWF (World Wildlife Fund). (2021). *Droughts in Central Europe: Impacts and Adaptation Strategies*. Brussels: WWF Europe.

WWF. (2021). *Managing Transboundary Waters: Challenges in Eastern Europe*. WWF Report.

WWF. (2021). *Transboundary Water Management in the Danube River Basin*. WWF Report.

5

Global Water Crisis–Water-Abundant Regions

While many regions worldwide experience severe water scarcity, there are also water-abundant regions with abundant water resources. These regions, however, still face significant challenges related to the equitable distribution and sustainable management of water resources. Countries like Canada, Brazil, and Russia are home to vast freshwater reserves. However, even these nations are not immune to the growing global water crisis, as pollution, infrastructure, and governance issues can still pose serious threats to water security.

5.1 South America

Water-abundant regions in the world are characterized by significant freshwater availability, often supported by large river systems, lakes, and high precipitation levels. South America, with its vast and diverse geography, is home to some of the most water-rich areas on the planet. The continent is particularly known for its immense river basins, such as the Amazon, La Plata, and Orinoco, which support extensive ecosystems, agriculture, and urban centers. Despite the abundance of water, water distribution, management, and infrastructure challenges remain prevalent, especially in regions experiencing rapid population growth or urbanization.

The Amazon Basin, which spans across nine countries in South America, is the largest and most biodiverse river basin in the world. It holds approximately 20% of the planet's freshwater supply and contributes significantly to the global hydrological cycle (Finer et al., 2018) The Amazon River itself is the largest by discharge, draining an area of around 7 million square kilometers. This immense water resource sustains not only the region's rich biodiversity but also millions of people who depend on the river for drinking water, agriculture, and transportation. However deforestation and industrial activities in the basin threaten the long-term sustainability of this water source, as changes in land use affect local water cycles and water quality (Barlow et al., 2019).

The La Plata Basin, covering parts of Argentina, Brazil, Paraguay, and Uruguay, is another key water-abundant region in South America. This basin is home to several major rivers, including the Paraná and Uruguay rivers, which provide water for agriculture, industry, and hydropower generation. Argentina's agricultural sector, particularly in the provinces of Córdoba and Santa Fe, relies heavily on irrigation from these rivers. However, water scarcity concerns are emerging due to competing demands between agriculture, urbanization, and environmental protection (Burlando et al., 2015). While the region generally experiences more stable rainfall patterns than other parts of the world, climate variability and increasing water demand may exacerbate tensions over water allocation in the future.

In contrast to the challenges faced by other water-scarce regions, the Orinoco Basin in Venezuela and Colombia is another example of a water-abundant area. The Orinoco River, one of the longest in South America, is a critical resource for local communities and supports large-scale agricultural production in both countries. However, the region also faces challenges related to water pollution from industrial and mining activities, which threaten water quality and limit access to clean water for rural populations (Silva et al., 2019). In addition, political instability in Venezuela has further complicated efforts to manage water resources effectively and equitably in the basin.

Despite their abundant water resources, these regions face significant water management and distribution challenges. Rapid urbanization, agricultural expansion, and environmental degradation are increasing pressure on water systems. While water availability is not necessarily the limiting factor in these regions, inadequate infrastructure, pollution, and competing demands between sectors create challenges for sustainable water use. Effective governance and cooperative management of shared water resources are essential for mitigating the risks posed by these challenges and ensuring equitable access to water for all.

5.1.1 Exemplary countries of abundance of water resources: Brazil (Amazon Basin), Colombia, Peru

South America is home to some of the most water-abundant countries in the world, primarily due to the vast and interconnected river basins that flow through the continent. Brazil, Colombia, and Peru are among the countries that benefit from these abundant water resources, largely through their ownership of the Amazon Basin, the largest river basin in the world. The availability of freshwater in these countries plays a crucial role in their economic development, biodiversity, and overall ecosystem health. However, this abundance is also accompanied by significant challenges in terms of water management, environmental protection, and equitable access.

Brazil stands out as the country with the largest share of the Amazon Basin, which contains approximately 20% of the world's total freshwater. The Amazon River itself is the largest river in the world by discharge, with a watershed that spans about 7 million square kilometers, covering a significant portion of Brazil's northern and western regions (Finer et al., 2018). This vast network of rivers and tributaries not only supports the livelihoods of millions of people but also sustains the world's most biodiverse ecosystem. The water resources in the Amazon are used for various purposes, including agriculture, drinking water, energy production through hydropower, and transportation. However, Brazil faces significant challenges in managing these resources sustainably. Deforestation, illegal logging, and agriculture in the Amazon region are leading to reduced water quality and disruptions in the local water cycle (Barlow et al., 2019). Climate change, particularly rising temperatures and altered rainfall patterns, is expected to exacerbate these issues, threatening the long-term stability of the water system and biodiversity in the region.

Colombia, which shares a portion of the Amazon Basin with Brazil, also benefits from the abundance of freshwater provided by its major river systems, including the Orinoco and Magdalena Rivers. These rivers provide essential water supplies for the country's agricultural, industrial, and urban sectors. The Magdalena River, in particular, is a critical waterway for transportation, connecting the country's interior with its Caribbean coast. However, Colombia faces challenges related to water pollution, especially in urban areas where untreated sewage and industrial waste often contaminate freshwater sources (Martínez et al., 2016). Additionally, the country is vulnerable to the impacts of climate variability, including prolonged droughts and heavy rains, which can disrupt the flow of rivers and reduce the availability of clean water for local populations. As Colombia continues to urbanize and develop its agricultural sector, sustainable water management practices will be crucial in ensuring the long-term availability of these vital resources.

Peru, located in the western part of South America, is also a key player in the continent's water landscape, with access to the Amazon Basin and its own significant river systems, such as the Marañón and Ucayali Rivers. These rivers contribute to the Amazon's flow and support the country's agriculture, which is heavily reliant on irrigation. Peru's highland regions, which are part of the Andean mountains, experience a mix of freshwater resources from glaciers, rivers, and rainfall. However, these water sources are increasingly under pressure due to climate change and the retreat of glaciers, which provide a critical water supply for both urban and rural areas (Bradley et al., 2021). The country's agricultural sector, particularly in the coastal regions, faces growing concerns over water scarcity due to competing demands from urban growth and industrial development. The Peruvian government has initiated various policies to improve water management, including large-scale infrastructure projects like reservoirs and irrigation systems to cope with water shortages, especially during dry spells (Pérez et al., 2018).

Other **South American countries** that benefit from substantial water resources include **Venezuela** and **Ecuador**. Venezuela's Orinoco River Basin provides a major freshwater source for the country, while Ecuador's high-altitude lakes and Andean river systems support its agriculture and hydropower generation. Both countries, like Brazil, face pressures on their water resources from industrial activity, pollution, and climate change, which threaten the availability of clean water for both ecosystems and human populations.

Despite the abundance of freshwater resources across these countries, there are several challenges related to water management. In many regions, water distribution is uneven, with some areas suffering from scarcity despite the overall abundance in the larger river systems. Urbanization, industrial growth, and agricultural demands place increasing pressure on water supplies, and climate change is expected to exacerbate these problems, particularly in areas that depend on glacier melt or seasonal rainfall. The key to ensuring the sustainability of South America's water resources lies in improving water governance, investing in infrastructure, and addressing environmental degradation to protect the rivers, lakes, and aquifers vital for current and future generations.

5.1.2 Abundant in renewable freshwater resources but faces challenges of distribution, pollution, and infrastructure issues

Although South America is abundant in renewable freshwater resources, the region faces significant challenges in managing, distributing, and ensuring access to these vital resources. The continent benefits from the vast Amazon Basin, extensive river systems, and abundant rainfall in many areas, positioning it as one of the world's water-rich regions (Finer et al., 2018). However, several factors—ranging from geographic disparities and uneven resource distribution to pollution and inadequate infrastructure—pose substantial obstacles to equitable access and sustainable management of these water resources.

5.1.2.1 Geographic distribution and water availability

One of the primary challenges in South America's water resources is the unequal distribution of freshwater. While countries like Brazil, Colombia, and Peru are blessed with extensive river systems, other regions, particularly in the Andean highlands and arid coastal areas, experience water scarcity despite the continent's overall abundance. For example, northern Chile and the coastal deserts in Peru rely on distant rivers, groundwater, or desalinated seawater due to limited local freshwater availability (Barlow et al., 2019). Similarly, the arid regions of Argentina and southern Brazil face water stress due to the lack of sufficient local freshwater sources, which are exacerbated by climate variability and seasonal droughts. This unequal distribution of water resources leads to significant regional disparities in access to clean water, resulting in tensions between urban and rural populations and between different economic sectors, such as agriculture, industry, and domestic use (Pérez et al., 2018).

5.1.2.2 Water pollution and environmental degradation

Despite abundant water, pollution remains a pervasive issue throughout South America. Rapid urbanization, industrialization, and agricultural expansion have contributed to the contamination of freshwater sources, especially in highly populated areas. For example, the Magdalena River in Colombia and the Paraná River, which flows through Brazil and Argentina, suffer from pollution due to untreated sewage, mining activities, and agricultural runoff (Martínez et al., 2016). Water pollution has detrimental effects on human health, causing waterborne diseases, and ecosystems that rely on clean water for survival. Water quality degradation also impacts local economies that depend on fishing and tourism, as well as agricultural practices that require clean irrigation water. In Brazil, deforestation in the Amazon region disrupts water cycles, leading to poorer water quality and more erratic rainfall patterns (Barlow et al., 2019).

5.1.2.3 Infrastructure challenges

Another significant issue is the region's inadequate water infrastructure, which hampers the efficient distribution and management of available water. Many South American countries face difficulties in providing clean water to all regions due to outdated or insufficient infrastructure. Rural areas, in particular, are often underserved, with limited access to potable water and sanitation systems. In urban areas, rapid population growth has outpaced the development of necessary infrastructure, resulting in unreliable water supply systems and frequent water shortages (Finer et al., 2018). Additionally, in countries like Venezuela and Bolivia, where political instability and economic challenges have weakened public institutions, the ability to maintain and expand water systems has been severely constrained. Poor infrastructure also exacerbates water loss, as leaks and inefficient distribution systems waste large quantities of water.

5.1.2.4 Climate change and future challenges

Compounding these challenges, climate change is expected to have profound effects on South America's water resources. Altered precipitation patterns, changing river flows, and the retreat of glaciers in the Andes threaten the stability of freshwater supply across the continent (Bradley et al., 2021). Glaciers, which act as natural reservoirs for many of South America's rivers, are shrinking rapidly, endangering the water supply for millions of people, especially in countries like Peru and Bolivia (Bradley et al., 2021). Furthermore, increasing temperatures and unpredictable rainfall will likely intensify existing water stress, particularly in vulnerable regions. These climate-related challenges will require South American countries to invest heavily in adaptive strategies, such as improving water storage, diversifying water sources, and investing in water-saving technologies.

5.2 Solutions and Future Directions

To address these challenges, South America must prioritize sustainable water management practices, such as better governance of water resources, investment in infrastructure, and enhanced pollution control. Integrated water resources management (IWRM) frameworks, which promote coordinated management of water, land, and related resources, can help ensure water is used equitably and efficiently (Pérez et al., 2018). Moreover, increasing investment in wastewater treatment, rainwater harvesting, and desalination technologies will be crucial in mitigating the impact of water pollution and scarcity. Additionally, efforts to combat deforestation in the Amazon and other critical ecosystems will be essential in preserving the natural water cycles that support the continent's water resources (Barlow et al., 2019). Ultimately, the future of South America's water resources will depend on the region's ability to adapt to changing conditions, manage resources sustainably, and ensure that the benefits of abundant freshwater are shared equitably across all populations.

5.3 Impact: Threats from Deforestation, Mining, and Urbanization Despite High Water Availability

Despite South America's high water availability, the region faces significant threats to its freshwater resources due to human activities such as deforestation, mining, and urbanization. These activities have profound environmental and social impacts, reducing the quality and availability of water across the continent.

5.3.1 Deforestation and its impact on water resources

One of the most significant threats to water availability in South America is deforestation, particularly in the Amazon Basin, one of the world's largest and most vital ecosystems. The Amazon is crucial in maintaining the regional and global water cycle, regulating rainfall, and replenishing river systems. However, deforestation—driven by agriculture, logging, and infrastructure development—has been rapidly diminishing the forest cover, disrupting local and regional hydrological cycles (Finer et al., 2018). Trees play a crucial role in absorbing and releasing water through evapotranspiration, and their removal leads to changes in rainfall patterns and reduced groundwater recharge. In Brazil, for example, the deforestation of the Amazon is contributing to droughts, reduced water flow in rivers, and water shortages in cities like São Paulo (Barlow et al., 2019). As forest cover diminishes, the region's freshwater systems become increasingly vulnerable to pollution, sedimentation, and fluctuating water levels.

5.3.2 Mining activities and water pollution

Mining, particularly in countries like Peru, Colombia, and Bolivia, has also contributed significantly to the degradation of freshwater resources in South America. The extraction of valuable minerals, such as gold, copper, and tin, often involves the use of toxic chemicals like mercury and cyanide, which are released into nearby water bodies, causing contamination and harming both human and aquatic life (Bebbington et al., 2017). In addition, the clearing of land for mining operations increases soil erosion and sedimentation in rivers, which can clog water sources and reduce water quality. In Peru's

Madre de Dios region, gold mining has caused widespread contamination of rivers, severely affecting local populations' access to clean water and exacerbating waterborne diseases (Luna et al., 2019). This pollution, compounded by limited regulatory enforcement, continues to pose a significant challenge to water security in mining areas.

5.3.3 Urbanization and its strain on water resources

Urbanization in South America has also intensified pressure on freshwater resources. Rapid population growth in cities like São Paulo, Buenos Aires, and Lima has led to increased water demand for domestic, industrial, and agricultural purposes. However, urbanization often outpaces the development of adequate water infrastructure, resulting in inefficient water distribution, leakage, and contamination. In many South American cities, water is sourced from rivers and aquifers that are already under stress from pollution, over-extraction, and reduced recharge rates due to urban sprawl (Pérez et al., 2018). Furthermore, sprawling urban development often leads to the destruction of natural water catchment areas and wetlands, which play a vital role in regulating water flow and maintaining water quality.

5.3.4 Impact on vulnerable populations

These threats are not only a concern for the environment but also for the vulnerable populations who depend on these water resources for their livelihoods. Indigenous communities, rural populations, and low-income urban dwellers often bear the brunt of water scarcity and pollution. In areas like the Peruvian Andes and Amazon, where water sources are already limited, deforestation and mining further exacerbate the difficulty of accessing clean water (Luna et al., 2019). Waterborne diseases, health problems associated with contaminated water, and the loss of traditional water sources have had devastating impacts on these communities, leading to long-term social and economic consequences.

Despite South America's high availability of freshwater resources, deforestation, mining, and urbanization present serious threats to the region's water security. To mitigate these threats, stronger policies and actions are needed to regulate and control deforestation, mining practices, and urban expansion. Protecting ecosystems such as the Amazon, improving water governance, and investing in infrastructure for sustainable water management are crucial steps in ensuring that the region's water resources continue to serve its growing population and ecosystems for generations to come.

5.4 Russia: The Largest Renewable Water Resources Globally, Concentrated in Siberia and Far from Population Centers

Russia, the largest country in the world, possesses vast renewable freshwater resources, with the majority of these resources concentrated in Siberia and the Far East. Russia's freshwater supply is one of the largest globally, yet its uneven distribution across the country presents significant challenges for managing water resources and ensuring equitable access. While these vast water reserves are abundant in remote regions, their distance from major population centers exacerbates the challenge of making this water accessible for domestic, industrial, and agricultural use.

5.4.1 Abundance of freshwater resources

Russia's renewable freshwater resources are immense, with an estimated 25% of the world's total freshwater found within its borders. These resources are primarily concentrated in Siberia, which is home to some of the world's largest freshwater lakes and rivers. The most notable example is Lake Baikal, located in southeastern Siberia. Lake Baikal is the world's deepest and oldest freshwater lake, containing approximately 20% of the world's unfrozen surface freshwater (Shvidenko et al., 2017). With a volume of about 23,600 cubic kilometers, it is not only a critical source of freshwater but also a vital ecological and cultural asset for Russia.

In addition to Lake Baikal, Russia is home to numerous large rivers, many of which flow through Siberia. The Lena, Yenisei, and Ob Rivers are three of the longest and largest rivers in the world, draining vast areas of Siberia and feeding into the Arctic Ocean. The Lena River, for example, stretches over 4,400 kilometers, making it one of the longest rivers in the world (Pavlenko et al., 2015). These rivers serve as vital freshwater sources for the regions they pass through, supporting ecosystems and providing water for human consumption and agriculture in remote areas. However, despite the size and abundance of these water bodies, the distance from urban areas such as Moscow and St. Petersburg means that much of this water remains inaccessible to the majority of Russia's population.

5.4.2 Distribution challenges and accessibility

The problem of water distribution in Russia is primarily due to the concentration of its freshwater resources in sparsely populated regions of Siberia.

According to some estimates, nearly 70% of Russia's total freshwater resources are found in the Siberian and Far Eastern regions, where fewer than 30% of the country's population resides (Frolov & Podkolzin, 2019). In contrast, the majority of Russia's population lives in the European part of the country, where water availability is much lower. This creates a significant challenge for water management and infrastructure, as water must be transported long distances to reach major urban centers, such as Moscow, Kazan, and St. Petersburg.

This spatial mismatch between water resources and population density also has environmental implications. Large-scale water infrastructure projects, such as reservoirs, pipelines, and aqueducts, are necessary to redistribute water from Siberia to more populated areas. However, these projects are expensive, logistically complex, and can have negative ecological consequences. For example, the construction of dams and reservoirs along major rivers like the Yenisei has led to the displacement of local communities and disruption of ecosystems (Pavlenko et al., 2015). Additionally, the long distances involved in transporting water can result in significant losses due to evaporation and leakage, further exacerbating the challenges of water management.

5.4.3 Water quality and pollution concerns

Although Russia is rich in freshwater, pollution remains a significant concern, particularly in industrial areas and major rivers like the Volga and the Don, which run through densely populated and industrialized regions. Pollution from agricultural runoff, industrial discharges, and untreated sewage has resulted in the degradation of water quality in many parts of the country. The Volga River, the longest river in Europe, has been particularly affected by pollution, with significant levels of chemical contaminants and heavy metals detected in its waters (Revich & Shaposhnikov, 2018). These contaminants threaten both the environment and human health, as the Volga is a crucial source of drinking water for millions of people.

In Siberia and the Far East, the relatively low population density means that freshwater resources are less affected by pollution compared to urban areas. However, rapid industrial development in these regions, especially in oil and gas extraction and mining, poses emerging threats to water quality. The exploitation of these resources, particularly in sensitive ecosystems like those surrounding Lake Baikal, has raised concerns about potential contamination and long-term environmental impacts (Frolov & Podkolzin, 2019). These concerns underscore the need for effective governance and water quality management across the country to safeguard its abundant freshwater resources.

Russia's freshwater resources are vast and globally significant, but the distribution of these resources creates challenges in terms of accessibility and sustainability. While Siberia and the Far East are home to some of the world's largest and most pristine freshwater bodies, the distance between these areas and population centers complicates water management efforts. The government faces the challenge of balancing the needs of urban populations with the environmental preservation of its water resources, particularly in light of increasing industrialization and climate change. Ensuring that Russia's vast water resources are effectively managed, protected, and made accessible to its population will require innovative infrastructure solutions and a commitment to sustainable water governance.

5.5 Canada: Plentiful Freshwater, Large Lake and River Systems, Relatively Low Water Stress

Canada is endowed with some of the world's most abundant freshwater resources, making it a country with relatively low water stress compared to many other nations. With vast lake and river systems, Canada holds approximately 20% of the world's total freshwater supply, a crucial resource for both its environment and economy. Despite this abundance, water management in Canada involves addressing regional disparities, pollution, and the impacts of climate change.

5.5.1 Abundance of freshwater resources

Canada is home to an extensive network of freshwater resources, including the world's largest freshwater lakes, such as Lake Superior, Lake Ontario, and Lake Erie, which form part of the Great Lakes system. The Great Lakes alone account for about 18% of the world's surface freshwater, serving as a critical water source for millions of people living in Canada and the United States. Additionally, the country's rivers, including the Mackenzie, Fraser, and St. Lawrence rivers, are vast and support a variety of ecosystems, industries, and agriculture. These rivers serve as essential channels for transportation, hydropower, and water supply, further emphasizing Canada's water abundance.

5.5.2 Regional disparities in water availability

While Canada as a whole enjoys ample freshwater resources, water availability is unevenly distributed across the country. The bulk of Canada's freshwater is found in the northern regions, with large rivers and lakes in areas such as the Canadian Shield and the boreal forest of the north. However, these areas are sparsely populated, and most of Canada's population is concentrated in the southern parts of the country, which experience drier conditions. For instance, southern Alberta and parts of British Columbia, despite having significant agricultural activity, face water scarcity issues due to relatively low annual precipitation. This geographical mismatch between water resources and population density poses challenges for water management, requiring infrastructure to transport water to more populated areas, especially in drought-prone regions.

5.5.3 Low water stress but growing concerns

Canada is fortunate in that it faces relatively low water stress compared to global standards, particularly in terms of per capita water availability. However, several factors are increasingly placing stress on local water resources. For example, the growing demands of agriculture, industry, and urban areas in water-scarce regions, such as the Prairies and southern British Columbia, can strain available supplies. Additionally, climate change is affecting precipitation patterns, with some areas experiencing longer periods of drought or more extreme weather events, which can exacerbate water scarcity in certain regions. This issue is particularly pressing in agricultural zones where water is essential for irrigation.

5.5.4 Water quality and pollution

While Canada enjoys plentiful freshwater, water quality is a concern in certain regions, particularly where industrial activities such as mining, forestry, and oil extraction take place. Pollutants from these industries, as well as untreated sewage and agricultural runoff, can contaminate water bodies, impacting local ecosystems and human health. For instance, the oil sands development in Alberta has led to concerns about the contamination of rivers and lakes with heavy metals and toxic chemicals, affecting water quality in the region. Additionally, urban areas across Canada face challenges related to the pollution of rivers and lakes, which necessitate investment in water treatment and pollution mitigation strategies.

5.5.5 Hydroelectric power and water use

Canada's abundance of freshwater also makes it a global leader in hydroelectric power production. The country generates more hydroelectricity than any other nation, accounting for approximately 60% of its total electricity production (International Energy Agency, 2020). This reliance on hydroelectric power has both positive and negative implications for water management. On the one hand, it underscores the importance of water for energy production. On the other hand, the construction of large dams and reservoirs can disrupt river ecosystems, affecting fish migration, water quality, and the flow of rivers downstream. Balancing the need for hydroelectric power with environmental sustainability remains a key challenge in Canada's water management policy.

Canada's wealth of freshwater resources places it in a favorable position when compared to many other countries, with relatively low water stress. However, the country faces challenges related to regional disparities in water availability, the impact of climate change, water pollution, and the need for infrastructure to manage its water resources effectively. Moving forward, sustainable water management practices will be essential to ensure that Canada's water resources are protected and equitably distributed, while also addressing the needs of growing urban centers and industries.

5.6 Impact- Russia And Canada: Largely Secure, But Environmental Conservation and Remote Access Remain Challenges

Both Russia and Canada are among the few countries with vast freshwater resources, positioning them as relatively secure in terms of water availability. However, despite their abundant freshwater reserves, both countries face significant challenges related to environmental conservation, the remote location of water sources, and the potential impacts of climate change on their ecosystems and water management systems.

5.6.1 Russia: Water security and environmental conservation challenges

Russia holds the largest renewable freshwater resources globally, with an extensive network of rivers, lakes, and reservoirs, particularly in Siberia, which is home to the world's deepest lake, Lake Baikal. Despite this abundance, much of Russia's freshwater is concentrated in remote and sparsely populated areas, far from the majority of its population centers, which are located in the western part of the country. This geographical mismatch poses logistical challenges for water distribution and access in densely populated urban areas.

Moreover, while Russia's vast water resources contribute to its water security, environmental conservation remains a significant issue. Pollution from industrial activities, such as mining, agriculture, and oil extraction, has compromised water quality in many regions. Notably, the extraction of natural resources from Siberia has led to contamination of rivers and lakes, raising concerns about water quality for local communities and ecosystems. The contamination of the Volga River, for example, has been attributed to industrial discharges and agricultural runoff, affecting both water availability and biodiversity in the region (Pallardy, 2018).

Furthermore, climate change poses an additional threat to Russia's water security. While Russia's freshwater resources are largely secure, the country faces the risk of altered precipitation patterns and reduced water availability in certain regions due to warming temperatures. The potential retreat of glaciers and changes to river flows could exacerbate water stress in regions that are already vulnerable, such as parts of the North Caucasus and southern Siberia.

5.6.2 Canada: Plentiful water, but remote access and environmental concerns

Similar to Russia, Canada enjoys abundant freshwater resources, including the Great Lakes and the Mackenzie River system. With approximately 20% of the world's freshwater supply, Canada has relatively low water stress. However, a key challenge lies in the geographical distribution of water resources. The majority of Canada's freshwater is located in the northern regions, which are sparsely populated, while the southern regions, home to a large portion of Canada's population, experience drier conditions, making water management more complex. This mismatch requires extensive infrastructure to transport water to urban areas and agricultural zones in the south (Sharma, 2018).

In addition to issues of remote access, Canada faces environmental conservation challenges related to water quality. Industrial pollution from mining, oil sands extraction, and forestry has led to contamination of freshwater systems in various regions, particularly in Alberta and British Columbia. For example, pollutants from the oil sands in Alberta have been linked to the contamination of water sources, raising concerns over the health of aquatic ecosystems and human populations that rely on these water bodies. While Canada has implemented policies to monitor and manage water quality, pollution from industrial activities continues to pose a significant threat to the long-term sustainability of its freshwater resources (Vörösmarty et al., 2018).

Climate change is another growing concern for Canada's water security. Although the country has abundant freshwater resources, the impacts of rising temperatures and shifting weather patterns could threaten water availability in certain regions. Prolonged droughts in southern regions, coupled with changes in precipitation patterns and increased evaporation, may lead to water shortages in areas heavily reliant on irrigation and agriculture, such as the Prairie provinces (Sharma, 2018). Additionally, the melting of glaciers and changes in river flows could affect the hydropower generation capacity in certain regions, which is a key source of energy for the country.

Both Russia and Canada enjoy abundant freshwater resources, contributing to their overall water security. However, challenges related to environmental conservation, remote access to water sources, and the potential impacts of climate change pose significant risks to their long-term water management strategies. Effective policies for pollution control, sustainable water use, and infrastructure development will be crucial in ensuring that these nations can maintain their water security in the face of evolving environmental and climatic challenges.

5.7 Northern Europe and the Nordic Region: Abundant Freshwater Resources and Distribution Maintenance

Northern Europe and the Nordic region, which includes countries like Sweden, Norway, Finland, Denmark, and Iceland, are fortunate to possess abundant freshwater resources, thanks to their vast river systems, lakes, and glaciers. These countries are among the top global leaders in terms of water availability per capita, with natural freshwater systems that provide ample water for domestic, agricultural, and industrial use. However, the challenges they face are not about scarcity, but rather about the effective distribution, maintenance, and management of these water resources, especially considering the region's geographic and climatic conditions.

5.7.1 Abundance of freshwater resources

The Nordic countries benefit from a unique hydrological setting, which includes an extensive network of rivers, lakes, and glaciers. Sweden, for example, has approximately 100,000 lakes, and Norway is home to numerous large fjords and freshwater lakes. Finland also has an impressive number of lakes—over 188,000—that significantly contribute to the country's freshwater supply. Iceland's glaciers and geothermal resources further contribute to its water abundance.

The Nordic countries are blessed with high rainfall, snowmelt, and a high degree of groundwater recharge, which ensures a constant supply of freshwater throughout the year. The region's vast forested landscapes also play an essential role in maintaining the water cycle, with forests acting as natural filters for water, reducing pollution, and maintaining water quality. These factors make Northern Europe one of the most water-abundant regions globally, with an average per capita water availability that is significantly higher than many other parts of the world.

5.7.2 Distribution and maintenance challenges

While freshwater resources are abundant in the Nordic region, the distribution and maintenance of these resources present several challenges, primarily due to the region's climate, geography, and population density.

Geographic Distribution and Population Centers: The Nordic countries, particularly Finland, Sweden, and Norway, have large, sparsely populated areas, with the majority of the population concentrated in a few urban centers. This geographic distribution means that while freshwater resources are abundant in certain regions, they must be effectively transported to cities and agricultural areas that are often located far from natural water sources. For instance, in Norway, the mountainous terrain and rugged landscapes can make infrastructure development for water distribution challenging.

Seasonal Variability: The seasonal variations in climate, such as freezing winters and wet summers, create significant shifts in the availability and flow of freshwater. For example, in the winter months, much of the water is locked in snow and ice, and it is only during the spring thaw that rivers and lakes experience increased water flow. Managing this seasonal variability requires efficient infrastructure to store and regulate water resources for use during drier months.

Glacial Melt and Water Quality: While glaciers provide a significant amount of freshwater in countries like Iceland and Norway, the melting of glaciers due to climate change poses a long-term threat to water availability. The increased runoff from glaciers can lead to higher levels of sediment in rivers, affecting water quality and making it harder to maintain the cleanliness of water supplies for human consumption and agriculture. Additionally, glacial meltwater often contains high levels of minerals and contaminants, which may need to be treated before use.

Pollution and Contamination: The Nordic countries, while known for their pristine environments, are not immune to pollution challenges. Industrial activities, urbanization, and agricultural runoff have led to contamination of freshwater resources in some areas. For instance, nitrates from fertilizers have polluted groundwater supplies in parts of Denmark and Sweden. While these countries have stringent environmental regulations, managing and preventing water pollution remains an ongoing challenge.

Climate Change: Climate change is likely to exacerbate water distribution challenges in the Nordic region. The region is experiencing rising temperatures, which are leading to more frequent and intense rainfall events, potentially resulting in flooding. At the same time, there are concerns that reduced snowfall could affect the seasonal replenishment of freshwater resources, impacting hydropower generation, a significant energy source in countries like Norway and Sweden. Additionally, changes in precipitation patterns and the timing of snowmelt could shift the availability of water, requiring adaptive water management strategies.

5.8 Management and Future Outlook

The Nordic countries have been proactive in managing their water resources, implementing policies and technologies to address both distribution and maintenance challenges. For example, Sweden, Finland, and Norway have established advanced water management systems that include water conservation programs, pollution control measures, and innovative technologies for wastewater treatment. Additionally, these countries are heavily invested in renewable energy sources, such as hydropower, which relies on the abundant freshwater from their rivers and lakes.

Despite these efforts, the region must continue to adapt to changing climatic conditions. Increased collaboration between governments, industries, and research institutions will be critical in ensuring that water resources are efficiently managed and protected for future generations. In the face of climate change, there is a growing need for sustainable practices that balance the demand for water with conservation and environmental protection.

The Nordic countries of Northern Europe are fortunate to have some of the most abundant freshwater resources globally, supported by an intricate network of rivers, lakes, and glaciers. However, challenges related to the distribution of these resources, particularly in sparsely populated areas, seasonal variability, pollution, and climate change, require ongoing attention. While the region has made significant strides in water management, adapting to the impacts of climate change and ensuring the long-term sustainability of water resources will remain a critical focus in the years to come.

5.8.1 Secure Water Supplies in Northern Europe and the Nordic Region

Northern Europe and the Nordic region, including countries like Sweden, Norway, Finland, Denmark, and Iceland, are known for their secure and reliable water supplies. These countries benefit from a combination of favorable geographical conditions, robust water management systems, and low population density, all of which contribute to a high level of water security.

5.8.1.1 Abundant precipitation

One of the key factors contributing to the region's secure water supplies is its relatively high levels of precipitation. The Nordic countries experience frequent rainfall and snowfall throughout the year, which helps to replenish rivers, lakes, and groundwater systems. In addition to rainfall, snowmelt during the spring provides a seasonal boost to river flows, which further sustains water resources. For instance, Norway and Sweden receive substantial precipitation from the North Atlantic storm tracks, ensuring that their water systems are continuously replenished.

This abundance of water means that there is generally a low risk of water scarcity in these countries, unlike many arid or semi-arid regions worldwide. Freshwater sources are plentiful and can easily meet the demands of both urban and rural populations. The availability of surface water from lakes and rivers is particularly significant in countries like Finland, which boasts one of the highest densities of lakes in the world.

5.8.1.2 Low population density

Another important factor contributing to water security in the Nordic region is the relatively low population density. This allows for more equitable distribution and greater availability of water per capita. While the total population of these countries is not small—Finland, Sweden, and Norway have populations ranging from 5 to 10 million people—their vast geographical areas result in lower concentrations of people in any given location. In fact, much of the population in these countries lives in urban centers, leaving rural and remote areas with abundant access to freshwater sources. The low population density reduces the pressure on water systems, ensuring that the available resources are sufficient to meet demand without the risks of over-extraction or pollution.

For example, in Norway, only about 5 million people are spread across an area roughly the size of the United Kingdom, which allows for significant natural reserves of freshwater that are not heavily used or polluted.

5.8.1.3 Strong water management systems

In addition to favorable natural conditions, the Nordic countries also benefit from highly effective water management systems that help ensure the sustainable use of water resources. These systems include sophisticated infrastructure for water treatment, distribution, and conservation, as well as robust policies for pollution control and conservation.

For example, Sweden, Norway, and Finland have long been leaders in water quality management, establishing stringent regulations to protect freshwater systems from contamination. These regulations help to ensure that water remains clean and safe for consumption, industrial use, and agricultural irrigation. In these countries, much of the water management infrastructure is modern, and public awareness regarding the importance of water conservation is high. Efficient water distribution systems help to prevent losses, ensuring that water resources are used optimally across urban, industrial, and agricultural sectors.

Moreover, the Nordic countries are also leaders in hydropower generation, which relies heavily on the region's abundant freshwater systems. Hydropower provides a significant portion of the energy used in these countries, further enhancing the integration of water resources into their national infrastructure.

5.8.1.4 Climate change adaptation

While climate change poses a global threat to water security, the Nordic region's water supplies are expected to remain secure in the short to medium term. The ongoing shift in weather patterns may lead to more intense rainfall events and altered snowmelt timing, but these changes are generally manageable due to the existing infrastructure and water management strategies.

That said, the region's focus on climate change adaptation is already evident. Countries in the Nordic region are investing in climate-resilient water systems, improving flood control mechanisms, and enhancing their capacity to manage extreme weather events. For instance, in Sweden and Norway, flood risk management systems have been put in place to address potential flooding from heavy rainfall and snowmelt in the coming years.

Northern Europe and the Nordic region enjoy secure water supplies due to their abundant precipitation, low population density, and advanced water management systems. The combination of these factors ensures that these countries face minimal risks related to water scarcity. With continued investment in infrastructure and climate adaptation strategies, the region is likely to maintain its strong water security in the face of future challenges. However, careful management remains essential to preserve these vital resources for future generations.

5.9 Impact of Minimal Water Stress in Northern Europe and the Nordic Region

The minimal water stress experienced by countries in Northern Europe and the Nordic region has several profound impacts, shaping their approach to water management, conservation, and environmental health. Unlike many regions facing critical water scarcity, these countries benefit from a stable and secure water supply, which allows them to focus on long-term sustainability and ecological preservation. However, this relative abundance of water resources does not negate the importance of responsible management. Rather, it enables these nations to prioritize environmental health, biodiversity conservation, and proactive climate adaptation efforts.

5.9.1 Focus on conservation and environmental health

With water stress being relatively low, countries in the Nordic region place significant emphasis on environmental health, which includes maintaining clean and sustainable water systems. Unlike regions where water shortages drive emergency responses, Northern Europe's governments can invest in projects that enhance water quality, prevent pollution, and protect aquatic ecosystems. This focus on conservation not only ensures the continued availability of high-quality water but also helps preserve the broader ecosystem, including wetlands, rivers, and lakes that play crucial roles in biodiversity.

For example, Sweden and Finland have implemented stringent policies to protect water quality, such as regulating industrial discharges and agricultural runoff, which can lead to eutrophication and other forms of water contamination. The goal is to protect aquatic life and ensure that freshwater ecosystems continue to thrive. These conservation efforts are supported by strong environmental laws, public engagement, and a high level of environmental awareness among the population.

5.9.2 Water-efficient practices and technologies

Although these countries are not facing immediate water scarcity, there is still a focus on water efficiency. In fact, the relative abundance of water has encouraged the development of cutting-edge technologies to improve water conservation and reduce waste. In countries like Denmark and Norway, water-efficient appliances, irrigation techniques, and industrial practices are common, helping to ensure that even abundant water resources are not misused.

For example, Denmark has made significant strides in reducing water waste by implementing smart metering systems, optimizing water treatment processes, and promoting rainwater harvesting. These efforts not only protect water resources but also contribute to the sustainability of urban and agricultural systems.

5.9.3 Climate change resilience and adaptation

Despite abundant water supplies, the Nordic region acknowledges that climate change presents challenges to even the most water-secure countries. The changing climate may lead to altered precipitation patterns, which could impact water availability, particularly in areas reliant on snowmelt. While the region's water systems are generally resilient, there is increasing investment in climate change adaptation strategies. This includes measures to manage flooding risks, address changing river flows, and protect water quality from extreme weather events.

Sweden, for example, has developed extensive flood risk management strategies that include building flood barriers, improving drainage systems, and restoring natural wetlands that can absorb excess water. Similarly, Norway's focus on preserving its freshwater systems from the impacts of climate change involves continuous monitoring and the development of adaptive water management systems.

5.9.4 Global leadership in sustainable water management

The minimal water stress in Northern Europe and the Nordic region has allowed these countries to take a leadership role in global water management. Through international collaborations and environmental diplomacy, countries like Sweden and Norway promote sustainable water use practices and advocate for the protection of freshwater resources worldwide. They actively participate in international agreements aimed at improving water governance, reducing water pollution, and ensuring equitable access to clean water globally.

For instance, Norway's support for water-related initiatives, including its contributions to the United Nations' Sustainable Development Goals

(SDGs) and various water-focused research organizations, highlights its commitment to global water sustainability. These countries not only focus on their own water security but also aim to share their expertise with regions facing more acute water stress.

5.9.5 Public awareness and education

Public awareness and education play a significant role in maintaining water security in these regions. With abundant resources, there is a focus on ensuring that citizens understand the importance of conserving water and protecting the environment. Environmental education initiatives often emphasize the role of water in ecosystems, the need for sustainable water management, and the impacts of human activity on water quality.

For example, Finland and Iceland promote educational programs that highlight water conservation practices, pollution prevention, and sustainable living. Schools and communities are regularly engaged in environmental campaigns, and these efforts foster a culture of water stewardship that benefits both present and future generations.

In Northern Europe and the Nordic region, the impact of minimal water stress is evident in the region's emphasis on conservation, water efficiency, and environmental health. While these countries benefit from an abundance of freshwater, they do not take this resource for granted. Rather, they invest in sustainable water management practices, climate adaptation strategies, and global leadership in water governance. As these countries continue to prioritize the protection of their water resources, they serve as models for other regions in promoting responsible water use and environmental sustainability.

References

Barbosa, A. F., Marengo, J. A., & Moreira, M. (2020). Climate change and its impacts on the water resources of Brazil. *Regional Environmental Change, 20*(4), 1–13. https://doi.org/10.1007/s10113-020-01615-3

Barlow, M., Clarke, T., & Smith, S. (2019). *Amazon deforestation and the impact on global water cycles.* Earth Institute.

Bebbington, A., Hinojosa, L., & Bury, J. (2017). *Mining and the water crisis in the Andes.* Latin American Perspectives, 44(1), 33–46.

Bradley, R., Vuille, M., & Finkel, M. (2021). *Glacier retreat and water resources in the Andes.* Journal of Hydrology, 594, 122–134.

Burlando, P., & Ramírez, A. (2015). Water management challenges in the La Plata Basin. *Water Resources Management*, 29(9), 3397–3410.

Finer, M., Jenkins, C. N., Pimm, S. L., Keane, B., & Ross, C. (2018). *The Amazon basin and its freshwater resources.* Global Change Biology, 24(3), 1391–1400.

Frolov, I., & Podkolzin, A. (2019). *Water resources and challenges of sustainable management in Russia.* Springer.

Gleick, P. H. (2000). The world's water volume 1: The biennial report on freshwater resources. *Island Press.*

Håkanson, L., & Boulton, M. (2015). *Water quality and quantity in Northern Europe: The impact of climate change and pollution.* Springer.

Håkanson, L., & Boulton, M. (2015). *Water quality and quantity in Northern Europe: The impact of climate change and pollution.* Springer.

International Energy Agency. (2020). *Hydropower.* https://www.iea.org/topics/hydropower Sharma, S. (2018). *Canada's water resources: An overview.* Springer.

Kuzmin, V. (2017). Water security in Russia: Challenges and prospects. *Water Resources, 44*(2), 138–145. https://doi.org/10.1134/S0097807817020036

Luna, F., Paredes, D., & Rodríguez, A. (2019). *Gold mining and water pollution in Peru's Madre de Dios region.* Environmental Science and Policy, 99, 68–74.

Martínez, D., Hernández, L., & Rodríguez, R. (2016). Pollution and water quality in the rivers of Colombia. *Environmental Management*, 58(5), 1253–1266.

Pallardy, R. (2018). *Russia: Environment.* Encyclopedia Britannica. https://www.britannica.com/place/Russia/Environment

Pavlenko, S. V., Ivanov, V. P., & Balandina, M. V. (2015). *Freshwater resources in Russia: Potential, usage, and challenges.* Russian Geology and Geophysics, 56(4), 369–375.

Pérez, F., Silva, L., & García, J. (2018). *Water scarcity and management in Peru: The role of irrigation systems.* Water Resources Management, 32(8), 2557–2573.

Revich, B., & Shaposhnikov, D. (2018). *Water quality and pollution in Russia: The Volga River case.* Environmental Health Perspectives, 126(11), 116002.

Rhein, M., & Moser, C. (2020). *Climate change and freshwater management in the Nordic region.* Nordic Council of Ministers.

Schindler, D. W. (2010). The Alberta oil sands and water quality: Lessons from a decade of research. *Environmental Science & Technology, 44*(21), 8090-8096. https://doi.org/10.1021/es101596f

Sharma, S. (2018). *Canada's water resources: An overview.* Springer

Shvidenko, A. Z., Nilsson, S., & McCallum, I. (2017). *Lake Baikal and its role in the global water cycle.* Global Environmental Change, 44, 25–34.

Silva, J. D., Carvajal, R., & Martínez, M. (2019). Pollution and management issues in the Orinoco Basin. *Environmental Management*, 63(4), 672–684.

Vörösmarty, C. J., McIntyre, P. B., & Gessner, M. O. (2018). *Global threats to human water security and river biodiversity*. Nature, 467(7315), 555–561.

6

The Disparities in Water Availability and Stress Worldwide

Understanding the significant variations in water availability and stress across the globe is of paramount importance. This comprehension is an academic exercise and a crucial tool for addressing water security challenges and ensuring sustainable water use for future generations. Some regions have abundant resources and low water stress, while others have severe scarcity. These disparities result from complex factors, including geographic location, climate conditions, population density, water management practices, and economic development.

6.1 Geographic and Climate Factors

Geography and climate play a major role in determining water availability. Tropical and temperate regions, such as South America, Southeast Asia, and Central Africa, tend to have abundant freshwater resources due to regular rainfall and large river systems. For example, the Amazon Basin in Brazil is home to one of Earth's largest and most water-rich regions, providing critical water supplies for both local populations and ecosystems (Barbosa et al., 2019). Similarly, Northern European countries like Sweden and Finland benefit from high precipitation levels and a favorable climate supporting abundant freshwater reserves (Håkanson & Boulton, 2015).

In contrast, arid and semi-arid regions such as the Middle East, North Africa, and South Asia face chronic water scarcity due to limited rainfall and high evaporation rates. Countries like Saudi Arabia, Yemen, and Egypt struggle with severe water stress as they rely heavily on groundwater and desalinated water, both costly and unsustainable in the long run (Falkenmark & Lannerstad, 2010). Climate change exacerbates these issues by altering precipitation patterns, increasing the frequency of droughts, and reducing the predictability of water supplies in regions already prone to water stress (IPCC, 2021).

DOI: 10.1201/97818743801061-6

6.2 Population and Economic Growth

The role of population density and economic development in the disparity of water availability cannot be overstated. High-density urban areas and industrialized regions exert significant pressure on available water resources, even in regions that would otherwise be considered water-abundant. For instance, India and China, with their large populations and rapidly growing economies, face significant water stress despite having considerable water resources. The over-extraction of rivers like the Ganges and the Yangtze, combined with pollution and inefficient water use, has led to widespread water scarcity, particularly in urban centers and agricultural regions.

On the other hand, regions with low population densities, such as Canada, Russia, and parts of Northern Europe, enjoy relatively low water stress due to their abundant freshwater resources and strong water management systems. In these areas, the combination of low demand and high availability allows for more effective conservation and management practices, ensuring water resources' sustainability (Sharma, 2018; Håkanson & Boulton, 2015).

6.3 Water Management and Infrastructure

The role of water management practices and infrastructure in effectively managing water resources cannot be overstated. Well-developed infrastructure, such as reservoirs, irrigation systems, and efficient water treatment plants, can help mitigate water scarcity in areas with limited natural water availability. However, many developing countries struggle with inadequate infrastructure, leading to significant water losses, poor water quality, and inefficient distribution systems.

For example, in Sub-Saharan Africa, inadequate infrastructure and poor governance contribute to physical and economic water scarcity. Despite having some freshwater resources, countries like Ethiopia, Somalia, and Sudan face widespread water shortages due to a lack of investment in water infrastructure, political instability, and inefficient management (Tsimpo et al., 2014). Conversely, countries such as Israel have invested heavily in innovative water management techniques, including drip irrigation and desalination, which have allowed them to overcome water scarcity challenges and increase their water security (Feitelson & Rosenthal, 2016).

6.4 Environmental and Ecological Considerations

Environmental and ecological factors also shape the disparities in water availability and stress. In water-abundant regions, such as the Amazon and

Congo Basin, deforestation, industrial pollution, and unregulated agricultural expansion threaten the integrity of freshwater ecosystems and the long-term sustainability of water resources. Similarly, over-extraction for agricultural irrigation, coupled with poor water management practices, can lead to salinization and the depletion of aquifers, critical water sources in arid regions (Liu et al., 2019).

In regions already facing water stress, such as parts of the Middle East and Central Asia, the degradation of freshwater ecosystems and the depletion of groundwater resources are compounded by increasing demands for water, leading to a vicious cycle of water scarcity and environmental degradation. The Aral Sea crisis, which has led to the shrinking of one of the world's largest lakes, is a prime example of how mismanagement of water resources and environmental factors can exacerbate water stress (Micklin, 2007).

6.5 The Impact of Climate Change

Climate change is a significant driver of disparities in water availability and stress. Rising global temperatures are affecting rainfall patterns, with some regions experiencing more intense and frequent droughts while others face heavier rainfall and flooding. In many areas, these changing climate conditions are further straining already stressed water systems, particularly in regions that rely on snowmelt or seasonal rainfall. In parts of Asia, including India and Pakistan, the melting of glaciers in the Himalayas has already begun to affect the flow of major rivers that provide water for millions of people (Hussain et al., 2019).

In contrast, regions already water-abundant, such as Northern Europe and parts of Canada, face relatively less impact from climate change in terms of water availability. However, they still need to adapt to changing weather patterns and extreme events. The increased frequency of extreme weather events, such as flooding, could pose new challenges for water management and infrastructure in these regions.

A wide range of factors, including geography, climate, population dynamics, water management practices, and the effects of climate change, shape disparities in water availability and stress. While some regions benefit from abundant water resources and relatively low stress, others face severe challenges due to insufficient infrastructure, population pressures, and environmental degradation. Understanding these disparities is crucial for developing targeted policies and strategies to ensure water security and sustainability, particularly in regions where water scarcity is a growing concern.

6.6 The Need for Tailored, Region-Specific Water Management Solutions

The need for tailored, region-specific water management solutions is crucial. It gives us hope that we can address water scarcity challenges effectively and ensure sustainable water use for future generations.

Water management is a critical issue that affects every part of the world. However, the challenges and needs vary significantly across regions due to climate, geography, economic conditions, and population density differences. As such, water management solutions cannot be one-size-fits-all; they must be tailored to the unique conditions of each region. Tailored, region-specific solutions are necessary to address the factors influencing water availability and quality, such as climate variability, economic development, infrastructure, and environmental degradation. Below, we discuss why region-specific solutions are essential and explore how they can be implemented in different parts of the world.

6.6.1 Geographic and climatic variability

Geography and climate are two of the most significant factors influencing water resources. Regions with abundant water, such as the Amazon Basin in South America or the Great Lakes region in North America, face different challenges compared to water-scarce areas like the Middle East, North Africa, and South Asia. In water-abundant areas, the primary concern may be pollution control, sustainable water use, and conservation. In contrast, in arid and semi-arid regions, the focus is often on water scarcity, efficient water use, and the construction of desalination plants (Falkenmark & Lannerstad, 2010).

For example, in the Middle East, where freshwater is scarce, countries like Saudi Arabia and Israel have turned to innovative water management solutions such as desalination and reusing treated wastewater. Meanwhile, in regions like Northern Europe, which has abundant freshwater resources, the challenges are more about maintaining water quality and minimizing pollution through effective waste management and industrial regulations (Sharma, 2018). Each region, therefore, requires different approaches based on local water availability, climatic conditions, and existing infrastructure.

6.6.2 Population and economic pressures

The level of water stress is also influenced by population density and economic activities. Highly urbanized regions with high population growth, such as India, China, and parts of Sub-Saharan Africa, face pressure on their water resources due to the combined demands of agriculture, industry, and domestic use. These areas require solutions to improve water use efficiency, reduce water losses, and invest in infrastructure to support growing populations. This might include expanding irrigation systems, developing water-efficient technologies, and implementing water pricing policies to encourage conservation (Falkenmark & Lannerstad, 2010).

On the other hand, sparsely populated and industrialized regions such as Canada or Russia may focus more on maintaining and protecting their abundant water resources to prevent future depletion and environmental degradation. In these regions, the challenge may be to ensure that water resources are not overexploited or polluted, especially in the face of expanding industries and urbanization (Sharma, 2018).

6.6.3 Environmental sustainability and conservation

Environmental concerns are a key driver in the development of region-specific water management strategies. For example, deforestation poses a significant threat to water quality and availability in areas like the Amazon Basin. Deforestation can disrupt the water cycle by reducing transpiration and cloud formation, leading to altered rainfall patterns and rivers and lakes drying up (Barbosa et al., 2019). Water management strategies must be closely linked with conservation efforts, including reforestation, forest protection, and sustainable land-use practices.

In contrast, regions such as the Mediterranean Basin, where water availability is often limited by droughts and prolonged dry seasons, require solutions that integrate water storage, conservation, and efficient irrigation techniques to reduce water wastage (Liu et al., 2019). With its mix of agricultural and tourism demands, the Mediterranean region requires innovative solutions like water-saving irrigation systems and policies regulating water use during peak tourist seasons to minimize pressure on water resources.

6.6.4 Technology and infrastructure development

Tailored solutions should also consider different regions' existing technological and infrastructural capacities. Advanced technologies, such as desalination plants, advanced wastewater treatment, and smart water metering, are more feasible in regions with sufficient financial resources and technical expertise, like parts of North America, Europe, and the Middle East. In these regions, the focus should be on optimizing water systems to improve efficiency and reduce waste (Feitelson & Rosenthal, 2016).

However, in developing regions with limited resources, investment in basic water infrastructure—such as wells, boreholes, and water storage tanks—is essential to meet the needs of rural communities and urban areas with inadequate access to clean water. In such areas, the focus should be on low-cost, scalable solutions that can be implemented with minimal investment while providing long-term benefits (Tsimpo et al., 2014).

6.6.5 Climate change and future challenges

The impacts of climate change will also necessitate tailored water management strategies in different regions. As the frequency of extreme weather events, such as floods and droughts, increases, regions must adapt their water management systems accordingly. For example, in areas prone to flooding, such as Southeast Asia, flood control systems and flood-resilient infrastructure will be essential to protect water resources and ensure freshwater availability during dry periods. In contrast, arid regions must improve drought management and expand access to alternative water sources like desalination or groundwater (IPCC, 2021).

Given the vast disparities in water availability and stress across the globe, it is clear that there is no one-size-fits-all solution to water management. Each region's unique geographic, climatic, economic, and social conditions require bespoke practical and sustainable solutions. Regional solutions, whether focused on conservation, infrastructure development, or technology adoption, must be designed with local realities in mind to ensure the long-term availability and health of water resources. As the world continues to grapple with the challenges of climate change and population growth, investing in region-specific water management strategies will be essential for achieving global water security.

6.7 The Importance of International Cooperation in Addressing Water Crises and Promoting Water Equity

Water is a fundamental resource that transcends national borders and affects the well-being of communities worldwide. As the global population grows, climate change accelerates, and water resources become increasingly stressed, addressing water crises and promoting water equity requires coordinated, international efforts. The significance of international cooperation in this regard cannot be overstated. Through collaboration, nations can address shared water challenges and ensure that water is distributed equitably, sustainably, and efficiently, benefiting all populations.

6.7.1 Shared water resources and transboundary water management

Many of the world's most critical water sources, such as major river basins like the Nile, the Mekong, and the Danube, are shared across national borders. These transboundary water bodies are essential for multiple countries' livelihoods, agriculture, and industries. Disputes over water use can exacerbate tensions, disrupt cooperation, and even lead to conflict. Hence, effective international cooperation is crucial in managing shared water resources. Agreements, such as the Indus Waters Treaty between India and Pakistan, provide frameworks for collaborative water management, ensuring that both nations share water equitably while preventing disputes (Rogers & Lydon, 2018). These agreements often involve joint monitoring, data-sharing, and conflict resolution mechanisms to manage water allocation and ensure sustainability.

Additionally, international cooperation is essential for managing the broader hydrological cycle. Climate change has led to unpredictable shifts in rainfall patterns, exacerbating existing challenges in water-scarce regions. By sharing hydrological data, research, and forecasting models, countries can better prepare for floods, droughts, and water shortages, enabling faster and more effective responses to emerging water crises (Wang, 2019).

6.7.2 Climate change and global water challenges

Climate change exacerbates global water crises, particularly in water-scarce regions. Rising temperatures, melting glaciers, and changing precipitation patterns significantly alter water availability, especially in vulnerable regions such as Sub-Saharan Africa and parts of Asia. Without international cooperation, countries would struggle to address the transnational nature of these challenges. For instance, the shrinking of glaciers in the Himalayas threatens the water supply for over a billion people in countries like India, China, and Nepal (Kaser et al., 2010). Regional collaborations, such as the South Asian Association for Regional Cooperation (SAARC), have the potential to tackle these problems more effectively by developing integrated water management strategies that account for changing hydrological patterns and support shared mitigation and adaptation initiatives.

Furthermore, climate-induced water stress will disproportionately affect poorer nations that lack the infrastructure and resources to cope with water scarcity. Wealthier nations, in collaboration with multilateral organizations like the United Nations and the World Bank, are responsible for assisting these nations in building water resilience through investments in infrastructure, water-efficient technologies, and education programs. The Paris Agreement underscores the importance of climate cooperation, and water management plays a central role in its broader environmental goals.

6.7.3 Water equity: Ensuring access to clean and safe water for all

Water equity is a fundamental aspect of international water cooperation. Over 2 billion people globally still lack access to safe drinking water (UNICEF & WHO, 2021). Economic, political, and social factors and geographic location often influence disparities in access to clean water. For example, low-income communities in rural areas, marginalized groups, and refugees often face barriers to accessing water due to inadequate infrastructure or discrimination.

International cooperation is key to addressing these inequities. Efforts like the United Nations' Sustainable Development Goal (SDG) 6, which aims to ensure universal access to clean water and sanitation, provide a global framework for equitable water access. By fostering cross-border partnerships and supporting international initiatives, countries can work together to expand infrastructure, improve water quality, and ensure water equity, especially in the most disadvantaged regions.

International funding mechanisms and collaboration with international organizations, such as the World Health Organization and UNICEF, are crucial in achieving water equity goals. These entities support countries in upgrading water and sanitation systems, training local communities in water management practices, and implementing policies that promote fair distribution.

6.7.4 Sustainable development and regional water cooperation

Sustainable development and economic growth are closely tied to effective water management. Water is essential for agriculture, energy production, industry, and human health. Economic development is stunted without reliable access to water, and human welfare suffers. This is especially true in regions heavily dependent on agriculture for their livelihoods, such as Sub-Saharan Africa, South Asia, and parts of Latin America. International cooperation can help countries in these regions build sustainable water management systems that increase agricultural productivity, protect ecosystems, and promote green technologies like water-efficient irrigation and rainwater harvesting.

Moreover, international cooperation can provide a platform for promoting integrated water resource management (IWRM), which encourages the sustainable use of water while balancing environmental, economic, and social needs. IWRM is critical for addressing water scarcity while maintaining ecosystem health, and successful regional projects like the Nile Basin Initiative showcase the potential for cooperation in developing equitable and sustainable water systems (Elhance, 2000).

6.7.5 Role of technology and innovation

Technological advancements in water management, such as desalination, wastewater recycling, and water-efficient agriculture techniques, are transforming how countries address water scarcity. International cooperation enables countries to share knowledge, technologies, and innovations that can help tackle water crises. For example, Israel's success in water management, particularly in desalination and wastewater reuse, has served as a model for other nations facing similar challenges. Countries can adopt best practices and customize solutions to their local contexts through international collaborations, accelerating their capacity to manage water resources more efficiently (Feitelson & Rosenthal, 2016).

Addressing the global water crisis and promoting water equity requires a concerted effort transcending national borders. International cooperation is essential for managing shared water resources, mitigating the impacts of climate change, ensuring equitable water access, and supporting sustainable development. By working together, countries can develop region-specific solutions that address local water challenges while contributing to global water security. Through knowledge sharing, financial assistance, and the adoption of innovative technologies, international partnerships can play a pivotal role in ensuring that water is available, accessible, and sustainable for all.

References

Barbosa, C., Monteiro, J. A., & Vieira, J. (2019). *Water resources in the Amazon Basin: Environmental threats and management strategies*. Springer.

Elhance, A. P. (2000). *Conflict and cooperation in international water management*. United Nations University Press.

Falkenmark, M., & Lannerstad, M. (2010). *The new green revolution: A new approach to water management for food production in the drylands*. Earthscan.

Feitelson, E., & Rosenthal, G. (2016). *Water management in Israel: Overcoming scarcity*. Palgrave Macmillan.

Håkanson, L., & Boulton, M. (2015). *Water quality and quantity in Northern Europe: The impact of climate change and pollution*. Springer.

Hussain, M., Zaman, Q., & Mujtaba, T. (2019). *Climate change and water scarcity in the Himalayas: The challenges and solutions*. Springer.

IPCC. (2021). *Climate change 2021: The physical science basis*. Cambridge University Press

Kaser, G., Grosshauser, M., & Marzeion, B. (2010). Contribution potential of glaciers to water availability in different climate regimes. *Proceedings of the National Academy of Sciences*, 107(47), 20210–20215.

Liu, J., Zhan, X., & Yang, H. (2019). *Sustainable water resources management in China: Policies and practices*. Wiley-Blackwell.

Micklin, P. (2007). *The Aral Sea disaster: A review of the implications for the region*. Environmental Management, 40(1), 20–31.

Rogers, P., & Lydon, P. (2018). *Water and conflict: Conflict and cooperation over transboundary water resources*. Cambridge University Press.

Sharma, S. (2018). *Canada's water resources: An overview*. Springer.

Tsimpo, C., Markandya, A., & Anoh, K. (2014). *Water scarcity and its impacts on African economies*. Cambridge University Press.

UNICEF & WHO. (2021). *Progress on household drinking water, sanitation, and hygiene 2000–2020: Five years into the SDGs*. UNICEF & WHO.

Wang, J. (2019). *Global water resources: Impact of climate change and water availability*. Springer.

7

Conclusion–Analysis of The Global Water Crisis

The book *"Analysis of The Global Water Crisis: Illustrating Regions with Extreme Water Scarcity to Those with Relative Abundance"* is a pioneering work that offers a comprehensive analysis of the global water crisis, presenting a clear and detailed picture of the disparities in water availability across different regions of the world. By categorizing regions based on their water stress levels—from those facing extreme water scarcity to those with abundant water resources—the book sheds light on the complexities of water distribution, accessibility, and the varying factors contributing to scarcity and abundance. This work explores the global situation and aims to bridge the knowledge gap and open new avenues for resolving or minimizing the freshwater crisis, thereby addressing the underlying environmental and humanitarian issues.

7.1 Overview of Global Freshwater Availability

The book's first section provides a foundational understanding of global freshwater availability, setting the stage for a deeper dive into the crisis. Freshwater, which constitutes only about 2.5% of the planet's total water, is unevenly distributed across the globe. While some regions are naturally endowed with abundant freshwater resources, others face extreme scarcity, struggling to meet the demands of their growing populations and expanding industries. The book highlights that less than 1% of this freshwater is easily accessible for human use, further complicating water distribution and accessibility. Various environmental, socio-economic, and geopolitical factors compound this uneven distribution.

The chapter elaborates on how the geographical location, climate patterns, and seasonal variations influence freshwater availability in different regions. For example, regions with a high concentration of arid or semi-arid climates,

DOI: 10.1201/97818743801061-7

such as the Middle East and North Africa (MENA), face significant challenges in accessing water due to limited precipitation, frequent droughts, and lack of natural water sources. Conversely, regions like South America, blessed with vast river systems such as the Amazon, have an abundance of water but still face challenges related to pollution, deforestation, and unequal distribution.

7.2 Factors Influencing Water Access

In addition to geographic and climatic factors, the book delves into several key drivers of water access and scarcity. The influence of population density is significant, particularly in rapidly growing urban areas where water demand is outstripping supply. Industrial and agricultural activities, particularly in regions with high water-intensive crops or large-scale industrial operations, are major contributors to the depletion of available freshwater. The book also addresses the role of infrastructure—particularly in developing countries—where inadequate water management systems and inefficient distribution networks exacerbate water scarcity. Furthermore, political and institutional frameworks—or the lack thereof—are critical to understanding how water resources are allocated, conserved, and managed across borders, with many countries failing to implement effective policies for water sustainability.

7.2.1 Extreme water scarcity regions

The book dedicates significant attention to the regions suffering from extreme water scarcity. The Middle East and North Africa (MENA) stand out as one of the most water-stressed regions, where the combination of low rainfall, high evaporation rates, and rapidly increasing demand due to urbanization and population growth makes water management particularly challenging. In South Asia, countries like India and Pakistan are struggling with severe water scarcity, largely due to the over-extraction of groundwater, the mismanagement of river basins, and the looming threat of glacial retreat. With its high vulnerability to climate change and lack of infrastructure, Sub-Saharan Africa is another region marked by water stress, where millions of people still lack access to clean and safe water. The book outlines how these regions are particularly vulnerable to climate change impacts and political instability, with conflicts over water resources frequently exacerbating the situation.

7.2.2 High to moderate water stress regions

The book also identifies regions with high to moderate water stress, where water management is critical to preventing future crises. Central Asia, for instance, suffers from the remnants of the Aral Sea disaster, with water scarcity and environmental degradation affecting millions. Similarly, Southern Europe and the Mediterranean Basin are experiencing increased water stress due to climate change, which has led to rising temperatures, changing precipitation patterns, and increasingly severe droughts. The Western United States, particularly California and Arizona, is another example of a region under high water stress, where seasonal droughts and high agricultural demands are depleting groundwater reserves and stressing water supply systems.

7.3 Regions with Moderate Water Security

The challenges are different but still significant in regions with moderate water security, such as Eastern and Central Europe and Southeast Asia. While these areas generally have better access to water, they face growing pressures from industrial and agricultural demands, pollution, and climate variability. For example, Southeast Asia, with its abundant rainfall and rivers like the Mekong and Chao Phraya, is not immune to water security issues. Rapid urbanization, deforestation, and over-extraction for agriculture threaten water quality and availability in these regions, requiring proactive water management and conservation policies.

7.4 Water-Abundant Regions

The book's final section focuses on regions with abundant water resources, such as South America, Russia, Canada, and Northern Europe. These areas are typically less stressed in terms of water availability, but challenges still exist. Despite the vast Amazon River and other major freshwater sources in South America, water pollution, deforestation, and inefficient infrastructure hinder equitable access to clean water. In Russia and Canada, while vast reserves of freshwater exist in the form of glaciers, lakes, and rivers, the remote location of many water sources, combined with environmental conservation concerns, presents significant logistical challenges. Northern Europe, with its abundant freshwater and efficient management systems, faces minimal water stress but is increasingly focused on conservation and sustainable water use.

This book serves as a critical resource for understanding the global water crisis by categorizing regions based on water scarcity and security. By examining the various factors influencing water access—such as climate, geography, population density, and infrastructure—the book provides a holistic view of different regions' challenges and offers insights into potential solutions. The hope is that this work will increase awareness of the disparities in water availability and inspire global cooperation and innovative solutions to address the freshwater crisis. By bridging the knowledge gap and offering practical strategies for better water management, the book aims to mitigate the environmental and humanitarian impacts of water scarcity.

7.5 Recommendations- the Ways to Minimize the Global Water Crisis

The book provides a comprehensive examination of water availability across different world regions. It highlights the urgent need for a multi-faceted, collaborative approach to addressing the global water crisis. This book underscores that while some regions grapple with severe water scarcity, others enjoy relative abundance, yet all face challenges that require tailored strategies. To ensure a sustainable and equitable future for water resources, it is essential to involve multiple agents of change, each playing a crucial role in the collective effort to improve water management and resilience. Below are detailed recommendations for key stakeholders based on the book's discussions:

7.5.1 Governments and policymakers

Governments at all levels—local, national, and regional—must take proactive steps to establish robust water governance frameworks. Policies need to be tailored to the specific water challenges of each region, whether it's managing extreme scarcity in the Middle East, addressing over-extraction in South Asia, or conserving abundant but remote resources in Canada and Russia. Key actions include:

7.5.1.1 Strengthening regulatory frameworks

Governments should implement and enforce regulations that protect water resources from pollution, overuse, and mismanagement. This includes setting limits on industrial and agricultural water use, protecting watersheds, and ensuring sustainable groundwater extraction.

7.5.1.2 Investing in Infrastructure

In regions facing water scarcity, there is an urgent need to invest in infrastructure for rainwater harvesting, desalination, wastewater recycling, and efficient irrigation systems. Investments should improve distribution networks for water-abundant regions to avoid wastage and ensure equitable access.

7.5.1.3 Climate adaptation strategies

Governments should integrate water management into broader climate adaptation plans, particularly in regions vulnerable to droughts, floods, and changing precipitation patterns.

7.5.2 International organizations and multilateral bodies

The book highlights the critical role of international cooperation in addressing transboundary water issues and promoting water equity. Organizations such as the United Nations, World Bank, and regional alliances can drive efforts to:

7.5.2.1 Facilitate cross-border water management

Many rivers and aquifers cross national boundaries, leading to potential conflicts over water usage. International bodies should mediate and support cooperative agreements to ensure fair and sustainable water sharing, particularly in regions like Central Asia and Sub-Saharan Africa.

7.5.2.2 Provide financial and technical support

Developing countries often lack the resources to implement advanced water management solutions. International organizations can offer funding, technical expertise, and capacity-building programs to enhance water security.

7.5.2.3 Set global standards and guidelines

Establishing international standards for water quality, conservation practices, and sustainable usage can help guide national policies and promote best practices.

7.5.3 Private sector and industry

Industries are among the largest water consumers, especially in agriculture, manufacturing, and energy production. The private sector must take responsibility for reducing its water footprint and contributing to sustainable water management. Recommendations include:

Adopting Water-Efficient Technologies: Companies should invest in water-saving technologies, such as precision agriculture, closed-loop systems, and water-efficient production processes, to reduce their impact on local water resources.

Corporate Social Responsibility (CSR) Initiatives: Businesses can play a significant role by engaging in CSR programs focusing on water conservation, community water projects, and education campaigns. This is particularly important in regions where industries contribute to water stress.

Supply Chain Management: Corporations should assess and manage the water risks in their supply chains, particularly in water-stressed regions like Southern Europe and Western United States, to ensure sustainable sourcing and production.

7.5.4 Non-governmental organizations (NGOs) and civil society

NGOs and civil society organizations are critical in advocating for sustainable water management, raising awareness, and mobilizing community action. Their efforts can include:

Community-Based Water Management: Empowering local communities to manage their water resources through training, capacity building, and the implementation of localized water-saving practices. This is especially effective in regions with moderate water security, such as Southeast Asia.

Advocacy and Awareness Campaigns: NGOs can advocate for stronger water policies, hold governments and corporations accountable, and raise public awareness about the importance of water conservation.

Support for Vulnerable Populations: Civil society organizations should focus on ensuring access to clean water for marginalized communities, particularly in extreme water scarcity regions like Sub-Saharan Africa, where access is a matter of survival.

7.5.5 Scientific community and researchers

Research and innovation are essential for developing new technologies and solutions to address water challenges. The scientific community can contribute by:

Advancing Water-Saving Technologies: Developing innovative solutions such as desalination, advanced water filtration, and smart irrigation systems to optimize water usage in both agriculture and industry.

Conducting Impact Assessments: Researchers should focus on assessing the impacts of climate change on water resources, helping policymakers to develop informed strategies for adaptation and resilience.

Promoting Interdisciplinary Research: Collaboration between hydrologists, economists, social scientists, and climate experts can lead to a more holistic understanding of water issues and solutions tailored to the needs of diverse regions.

7.5.6 Local communities and individuals

Ultimately, the success of water management initiatives depends on the actions of individuals and communities. Empowering local populations to take an active role in water conservation can lead to sustainable change:

Water Conservation Practices: Encouraging the adoption of water-saving practices at the household level, such as rainwater harvesting, greywater recycling, and reducing water waste, is crucial for regions facing moderate to high water stress.

Community Engagement: Local leaders and community groups should collaborate to protect local water sources, manage shared resources, and advocate for sustainable practices.

Education and Behavior Change: Raising awareness about the value of water and the need for conservation can lead to behavioral changes that collectively contribute to water sustainability.

7.6 A Collaborative Path Forward

The global water crisis is a complex issue that requires a multifaceted and collaborative approach. As highlighted in *"Analysis of The Global Water Crisis"*, addressing water scarcity and ensuring sustainable management requires the combined efforts of governments, international organizations, industries, NGOs, scientists, and local communities. By leveraging the strengths and resources of these diverse agents of change, we can move towards a future where water is managed equitably and sustainably, mitigating the humanitarian and environmental crises linked to water scarcity. This book aims to bridge the knowledge gap, inspire action, and offer practical solutions that cater to the unique challenges of each region, ultimately contributing to global water security and resilience.

7.7 Future Directions of Research on the Global Water Crisis

Addressing the global water crisis requires an integrated and forward-thinking research agenda that spans all water ecosystems, from freshwater rivers and lakes to coastal and marine environments. Given the increasing pressure on water resources due to climate change, population growth, industrialization, and unsustainable consumption, future research must prioritize holistic, interdisciplinary approaches adaptable to diverse regional contexts. Below are key areas for future research, emphasizing the need for a comprehensive understanding of the dynamics within different water ecosystems and innovative strategies to ensure water sustainability.

7.7.1 Integrated water resources management (IWRM) across all ecosystems

A critical direction for future research is the expansion of Integrated Water Resources Management (IWRM) principles to encompass all water ecosystems. Traditionally, IWRM has focused on freshwater resources, but the interconnectedness of various water systems—rivers, lakes, wetlands, estuaries, and oceans—necessitates a broader approach. Research should explore how water management strategies can be integrated across different ecosystems to optimize resource use, reduce conflicts, and enhance resilience to environmental changes. For example, studies on river-ocean interactions can help understand nutrient flows, sediment transport, and their impacts on coastal water quality (Falkenmark & Rockström, 2010).

7.7.2 Climate change adaptation and resilience in water ecosystems

With climate change altering precipitation patterns, increasing the frequency of extreme weather events, and accelerating glacial melt, there is an urgent need for research on the adaptation and resilience of water ecosystems. Future studies should focus on:

Glacier-fed Rivers: Understanding the long-term impacts of glacial retreat on water availability, particularly in regions dependent on seasonal meltwater, such as the Andes and the Himalayas (Pritchard, 2019).

Wetlands and Floodplains: Investigating the role of wetlands in flood mitigation, carbon sequestration, and biodiversity conservation, and how they can be leveraged for climate adaptation.

Coastal Ecosystems: Exploring the effects of sea-level rise on freshwater aquifers, estuaries, and coastal communities, and identifying adaptive strategies for managing saltwater intrusion.

7.7.3 Innovative technologies for water conservation and reuse

Technological innovation is crucial for addressing water scarcity and enhancing water security across ecosystems. Future research should prioritize the development of advanced technologies for:

Desalination and Water Reuse: Exploring cost-effective and energy-efficient desalination techniques, especially for water-stressed coastal regions, and enhancing wastewater recycling systems to reduce the pressure on freshwater sources (Elimelech & Phillip, 2011).

Smart Water Management: Implementing digital technologies such as IoT sensors, artificial intelligence, and blockchain for real-time monitoring of water quality, efficient distribution, and leak detection in urban and agricultural settings.

Nature-based Solutions: Promoting the use of nature-based solutions, such as constructed wetlands, green infrastructure, and ecosystem restoration, to enhance water purification, reduce runoff, and restore degraded water bodies.

7.7.4 Socioeconomic and policy dimensions of water security

The global water crisis is as much a social and economic issue as it is an environmental one. Future research should address the policy, governance, and socioeconomic aspects of water management by focusing on:

Equitable Water Allocation: Investigating how water policies can be designed to ensure fair distribution among agricultural, industrial, and domestic users, particularly in regions experiencing high water stress, such as Central Asia and the Middle East (Gleick, 2014).

Transboundary Water Governance: Exploring mechanisms for cooperative water management in transboundary river basins, with a focus on conflict resolution, resource sharing, and joint infrastructure development.

Impact of Water Pricing and Market Mechanisms: Evaluating the effectiveness of water pricing, trading schemes, and subsidies in promoting sustainable water use, especially in regions like Australia's Murray-Darling Basin.

7.7.5 Conservation of freshwater biodiversity

Protecting freshwater biodiversity is crucial for maintaining ecosystem services, yet freshwater species are among the most threatened due to habitat loss, pollution, and over-extraction. Future research should focus on:

Species Conservation Strategies: Developing strategies to protect endangered freshwater species and their habitats, including riverine and lacustrine ecosystems.

Impact of Pollutants: Studying the effects of emerging contaminants, such as microplastics and pharmaceuticals, on aquatic life and ecosystem health, with an emphasis on developing bioremediation techniques.

Biodiversity Offsets and Restoration: Exploring the use of biodiversity offsets, ecological engineering, and restoration projects to mitigate the impacts of infrastructure development and agricultural expansion on freshwater habitats.

7.7.6 Holistic management of urban water systems

Urbanization presents unique challenges for water management, with cities facing issues like stormwater runoff, pollution, and aging infrastructure. Future research should prioritize:

Water-Sensitive Urban Design: Promoting research on integrating green infrastructure, such as green roofs, permeable pavements, and rain gardens, to enhance urban water resilience and reduce the urban heat island effect.

Sustainable Urban Drainage Systems (SUDS): Investigating how cities can adapt to increased rainfall variability and extreme weather events by implementing SUDS to manage stormwater and prevent flooding (Fletcher et al., 2015).

Urban Groundwater Management: Addressing the over-extraction of groundwater in urban areas, focusing on sustainable recharge methods and preventing contamination from industrial and domestic sources.

Conclusion

The future of research on the global water crisis must adopt a comprehensive, interdisciplinary approach that addresses the complexities of all water ecosystems. By focusing on the integration of technological advancements, climate adaptation strategies, policy reforms, and community engagement, research can provide innovative solutions to ensure sustainable water management and equity. As we move forward, fostering international collaboration and leveraging scientific knowledge will be critical to overcoming the challenges of water scarcity and ensuring a resilient water future for all.

References

Elimelech, M., & Phillip, W. A. (2011). The Future of Seawater Desalination: Energy, Technology, and the Environment. *Science*, 333(6043), 712–717.

Falkenmark, M., & Rockström, J. (2010). *Balancing Water for Humans and Nature: The New Approach in Ecohydrology*. Earthscan.

Fletcher, T. D., Shuster, W., Hunt, W. F., et al. (2015). SUDS, LID, BMPs, WSUD and more – The evolution and application of terminology surrounding urban drainage. *Urban Water Journal*, 12(7), 525–542.

Gleick, P. H. (2014). *The World's Water Volume 8: The Biennial Report on Freshwater Resources*. Island Press.

Pritchard, H. D. (2019). Asia's shrinking glaciers protect large populations from drought stress. *Nature*, 569(7758), 649–654.

Anex 1. Binding and Nonbinding Treaties, Conventions, and Agreements

Various binding and non-binding treaties, conventions, and agreements have addressed transboundary water issues. Below is a comprehensive list, along with their dates and the countries involved:

Binding Treaties, Conventions, and Agreements

1. **Convention on the Protection and Use of Transboundary Watercourses and International Lakes (Helsinki Water Convention)**
   - **Date**: 1992
   - **Countries Involved**: United Nations Economic Commission for Europe (UNECE) member countries. Opened for global accession in 2016, allowing non-European countries to join.
2. **United Nations Convention on the Law of the Non-Navigational Uses of International Watercourses (UN Watercourses Convention)**
   - **Date**: Adopted in 1997, entered into force in 2014
   - **Countries Involved**: Ratified by 37 countries (as of 2024), including France, Germany, South Africa, and Iraq.
3. **Revised Protocol on Shared Watercourses in the Southern African Development Community (SADC) Region**
   - **Date**: Signed in 2000, revised from the original 1995 protocol
   - **Countries Involved**: 16 SADC member countries, including South Africa, Botswana, Mozambique, and Zambia.

4. **Agreement on the Cooperation for the Sustainable Development of the Mekong River Basin (Mekong Agreement)**
   - **Date**: 1995
   - **Countries Involved**: Cambodia, Laos, Thailand, and Vietnam.
5. **Treaty Between India and Pakistan on the Utilization of the Waters of the Indus River (Indus Waters Treaty)**
   - **Date**: 1960
   - **Countries Involved**: India and Pakistan.
6. **Convention on Cooperation for the Protection and Sustainable Use of the River Danube (Danube River Protection Convention)**
   - **Date**: 1994
   - **Countries Involved**: 14 Danube Basin countries, including Germany, Austria, Hungary, and Romania.
7. **Convention on the Protection of the Rhine (Rhine Protection Convention)**
   - **Date**: 1999
   - **Countries Involved**: Switzerland, Germany, France, Luxembourg, and the Netherlands.
8. **Treaty Between the United States of America and Mexico on the Utilization of the Waters of the Colorado and Tijuana Rivers and of the Rio Grande**
   - **Date**: 1944
   - **Countries Involved**: United States and Mexico.
9. **Agreement on Cooperation for the Sustainable Development of the Senegal River Basin (Senegal River Basin Agreement)**
   - **Date**: 1972
   - **Countries Involved**: Mali, Mauritania, and Senegal.
10. **Niger Basin Water Charter**
    - **Date**: 2008
    - **Countries Involved**: Niger Basin Authority (NBA) member states, including Nigeria, Niger, and Mali.

11. **Columbia River Treaty**
    - **Date**: 1964
    - **Countries Involved**: United States and Canada.
12. **Zambezi Watercourse Commission (ZAMCOM) Agreement**
    - **Date**: 2004
    - **Countries Involved**: Angola, Botswana, Malawi, Mozambique, Namibia, Tanzania, Zambia, and Zimbabwe.
13. **Treaty of Peace between Jordan and Israel (Water Sharing Agreement as part of the treaty)**
    - **Date**: 1994
    - **Countries Involved**: Jordan and Israel.
14. **Guarani Aquifer Agreement**
    - **Date**: 2010
    - **Countries Involved**: Argentina, Brazil, Paraguay, and Uruguay.
15. **Agreement on the Establishment of the Volta Basin Authority**
    - **Date**: 2007
    - **Countries Involved**: Benin, Burkina Faso, Côte d'Ivoire, Ghana, Mali, and Togo.
16. **Kishenganga Agreement (Partial Award by Permanent Court of Arbitration)**
    - **Date**: 2013
    - **Countries Involved**: India and Pakistan.

Non-Binding Agreements and Frameworks

1. **Berlin Rules on Water Resources**
    - **Date**: 2004
    - **Countries Involved**: International set of guidelines adopted by the International Law Association, not legally binding but influential in water management.

2. **UNESCO Internationally Shared Aquifer Resources Management (ISARM) Initiative**
   - **Date**: Initiated in 2000
   - **Countries Involved**: Global initiative involving countries with shared aquifers.

3. **Bellagio Draft Treaty on Transboundary Groundwaters**
   - **Date**: Drafted in 1989
   - **Countries Involved**: Not a formal treaty but serves as a model for potential agreements.

4. **Global Environment Facility (GEF) International Waters Program**
   - **Date**: Ongoing since 1991
   - **Countries Involved**: Multiple countries in various international water bodies.

5. **EU Water Framework Directive**
   - **Date**: 2000
   - **Countries Involved**: European Union member states.

6. **Regional Strategic Action Plan on Integrated Water Resources Development and Management in the Southern African Development Community (SADC)**
   - **Date**: Adopted in 1998, updated periodically
   - **Countries Involved**: SADC member states.

7. **World Water Forum Declarations**
   - **Date**: Every three years since 1997
   - **Countries Involved**: Over 100 countries participate in forums.

Notable Historical Treaties

1. **Treaty of Albufeira (Iberian Water Convention)**
   - **Date**: 1998
   - **Countries Involved**: Spain and Portugal.

2. **Boundary Waters Treaty**
   - **Date**: 1909
   - **Countries Involved**: United States and Canada.
3. **Agreement between Egypt and Sudan on the Nile Waters**
   - **Date**: 1959
   - **Countries Involved**: Egypt and Sudan.

These agreements, conventions, and treaties are critical in addressing shared water resources, promoting cooperation, and preventing conflicts over water. While some are legally binding, others are guiding frameworks to promote sustainable and equitable use of transboundary watercourses.

Index

Words / Phrases Definitions–Analysis of The Global Water Crisis

Access to Clean Water: Refers to the availability and safe delivery of water that meets health standards, providing individuals and communities with the resources needed for drinking, sanitation, and hygiene. Access is crucial for public health and is often measured by the distance to the nearest safe water source, the reliability of supply, and the affordability of the service.

Adaptive Water Management: Involves adjusting water management strategies to cope with changing environmental conditions, including climate change, population growth, and shifting economic needs. It emphasizes flexibility, monitoring, and using science-based knowledge to guide decision-making in response to uncertainty and evolving water needs.

Agricultural Demand: Refers to the water required for irrigation, livestock, and other agricultural activities. Agriculture is a significant water consumer globally, and the demand for water in this sector is growing due to population increases, changing diets, and climate change, which affect water availability and agricultural productivity.

Aquatic Ecosystems: These are ecosystems that depend on water for survival, including rivers, lakes, wetlands, and oceans. They provide numerous ecological services such as water filtration, carbon sequestration, habitat for biodiversity, and regulation of local climate, all of which are essential for human life and environmental health.

Aquifer Depletion: Occurs when water is extracted from an aquifer (a layer of underground water-bearing rock) faster than it can be replenished by natural processes like rainfall. Over-extraction can lead to the drying up of wells, the deterioration of water quality, and land subsidence, making it a critical issue in regions reliant on groundwater.

Arid Climate: A climate characterized by low rainfall and high evaporation rates, resulting in a scarcity of water resources. Arid regions are often prone to droughts and struggle with water scarcity, impacting agricultural production, water availability, and human livelihoods.

Blue Economy: Refers to the sustainable use of ocean and water resources for economic growth, improved livelihoods, and jobs while preserving the health of aquatic ecosystems. It includes sectors like fisheries, tourism, renewable energy (e.g., offshore wind), and water management, all of which depend on healthy, sustainable water systems.

Brine Discharge: The release of highly concentrated salt water, often as a byproduct of desalination or industrial processes, into water bodies like oceans or rivers. This can negatively impact aquatic ecosystems by raising salinity levels, harming marine life, and disturbing natural habitats.

Canada Water Security: Refers to the protection and sustainable management of Canada's vast freshwater resources. Although Canada has abundant water resources, challenges like pollution, over-extraction, and climate change pose risks to water security, especially in regions with growing populations and industrial demands.

Climate Change: A long-term alteration of temperature and typical weather patterns in a place is primarily driven by human activities, especially the burning of fossil fuels. Climate change leads to more extreme weather events, rising sea levels, shifts in ecosystems, and affects water availability, which exacerbates water stress.

Climate Resilience: The ability of individuals, communities, and ecosystems to anticipate, prepare for, respond to, and recover from the impacts of climate change. In water management, resilience involves adapting water systems to withstand extreme weather, such as floods, droughts, and changing water availability.

Community-Based Water Management: A water management approach that emphasizes the involvement of local communities in decision-making, planning, and implementing water projects. It seeks to ensure that water resources are managed sustainably and equitably, taking into account the needs and rights of local populations.

Conflict Over African Water Resources: Refers to disputes between African nations or communities over the access, use, and management of shared water resources, particularly transboundary rivers and lakes. Such conflicts can arise due to differing national interests, political tensions, and the growing scarcity of water resources across the continent.

Conservation Strategies: Approaches designed to preserve and protect water resources and aquatic ecosystems, ensuring their sustainability. These strategies may include promoting water-efficient practices, reducing pollution, protecting wetlands, reforestation, and restoring degraded ecosystems.

Cross-border Water Tensions: Disagreements or conflicts between countries or regions sharing a water resource, such as a river or watershed. These tensions are often exacerbated by differing national priorities, population pressures, and competing uses of water, such as for agriculture, industry, and consumption.

Deforestation and Water Scarcity: The destruction of forests can lead to reduced water availability and increased water scarcity. Forests play a critical role in the water cycle by maintaining soil moisture, preventing erosion, and regulating water flows. Deforestation disrupts these processes, exacerbating droughts and water shortages.

Demand-Side Water Management: Involves strategies aimed at reducing water consumption, such as improving water efficiency, promoting conservation practices, and encouraging responsible water use. It contrasts with supply-side management, which focuses on increasing available water resources.

Desalination: The process of removing salt and other impurities from seawater to produce fresh water. Desalination is increasingly used in water-scarce regions but is energy-intensive, expensive, and can have environmental impacts such as brine discharge into marine ecosystems.

Desertification: The process by which fertile land becomes desert, typically as a result of drought, deforestation, or inappropriate agricultural practices. Desertification leads to a loss of biodiversity, reduced agricultural productivity, and increased water scarcity.

Drought Management: The implementation of strategies and actions to mitigate the effects of drought on water resources, agriculture, and human populations. This can include water conservation, water rationing, improved irrigation techniques, and drought-resistant crops.

Drought Resilience: The capacity of communities, ecosystems, and economies to recover from or adapt to the impacts of drought. It involves proactive measures like water conservation, developing drought-resistant infrastructure, and fostering adaptive agricultural practices.

Economic Impact of Water Shortages: Refers to the negative effects of water scarcity on economies, including reduced agricultural productivity, increased costs of water supply, diminished industrial output, and the potential for job losses and economic instability in water-stressed regions.

Ecosystem Services: The benefits that humans derive from ecosystems, including clean water, food, flood regulation, and biodiversity. Healthy aquatic ecosystems provide vital services such as water purification, flood control, and carbon sequestration.

Environmental Degradation: The deterioration of the environment through activities such as deforestation, pollution, and over-extraction of natural resources. In the context of water, environmental degradation can result in reduced water quality, habitat destruction, and biodiversity loss.

Environmental Factors: Elements of the natural environment that influence human and ecological health, including air quality, water availability, temperature, and the presence of pollutants. These factors play a crucial role in shaping water availability and quality.

Extreme Water Scarcity: Occurs when water resources are insufficient to meet the basic needs of populations, leading to serious consequences for health, agriculture, and industry. It is often exacerbated by climate change, over-extraction, and pollution.

Flood Management: Involves planning, policies, and technologies designed to reduce the risks associated with flooding, such as flood control systems, improved drainage, and early warning systems. Effective flood management is crucial for protecting human lives, infrastructure, and ecosystems.

Freshwater Availability: Refers to the amount of water that is accessible for human use, including drinking, sanitation, agriculture, and industry. Freshwater availability varies by region and is influenced by factors such as precipitation, population density, and infrastructure.

Freshwater Conservation: The protection and sustainable use of freshwater resources to ensure that they remain available for future generations. Conservation strategies include reducing water waste, protecting natural water sources, and improving the efficiency of water use in agriculture and industry.

Freshwater Pollution: The contamination of freshwater resources by pollutants such as chemicals, heavy metals, agricultural runoff, and sewage. Pollution degrades water quality, making it unsafe for consumption and harmful to aquatic ecosystems.

Freshwater Resources: Natural sources of water that are available for human use, such as rivers, lakes, and groundwater. These resources are crucial for drinking, agriculture, sanitation, and industrial processes. The sustainable management of freshwater resources is essential to ensuring long-term water security.

Global Water Crisis: Refers to the growing challenges related to the availability, quality, and management of water resources worldwide. The crisis is driven by factors such as population growth, pollution, climate change, mismanagement of resources, and over-extraction. It results in water scarcity, reduced access to clean water, and increasing tensions over water resources, with significant impacts on public health, food security, and ecosystems.

Governance Challenges in Water Management: These refer to the difficulties in effectively managing water resources, often exacerbated by political, economic, and institutional factors. Challenges include fragmented governance structures, lack of coordination across sectors and borders, corruption, inequitable distribution of resources, and insufficient legal frameworks to address water-related issues. Effective governance is crucial to ensuring equitable and sustainable water use.

Groundwater Depletion: Occurs when groundwater is extracted at a faster rate than it is naturally replenished. This leads to a decline in the water table, making wells dry up and reducing the availability of water for drinking, irrigation, and industrial use. Groundwater depletion is a significant concern in many arid and semi-arid regions and can result in ecological damage and increased costs for water extraction.

Groundwater Management: Involves the sustainable use and protection of groundwater resources. This includes measures like controlling extraction rates, monitoring aquifers, preventing contamination, and implementing policies for the equitable distribution of groundwater. Effective groundwater management is essential for ensuring long-term water security, especially in regions that rely heavily on underground water sources.

High Evaporation Rates: Occurs in regions with high temperatures, low humidity, and intense sunlight, where water quickly evaporates from surface water bodies and soils. High evaporation rates can exacerbate water scarcity, reduce water availability for agriculture, and affect the sustainability of reservoirs and rivers. This is especially problematic in arid and semi-arid regions.

Holistic Water Management: A comprehensive approach to water management that considers the interconnectedness of water systems, human activities, ecosystems, and the broader environment. It emphasizes the integration of water-related policies, economic considerations, and social equity, aiming to balance water supply and demand while protecting the health of water ecosystems and communities.

Hydroelectric Power: The generation of electricity from flowing or falling water. Hydroelectric power plants harness the energy of water to turn turbines and generate electricity. While it provides a renewable energy source, it can also have significant environmental and social impacts, including disruption of local ecosystems, displacement of communities, and water resource management challenges.

Hydrology: The scientific study of water, including its distribution, movement, and properties across the Earth. Hydrology is crucial for understanding the water cycle, managing water resources, and predicting and mitigating water-related disasters such as floods, droughts, and water scarcity.

Hydropolitics: Refers to the political and strategic issues related to the control, allocation, and use of shared water resources. Hydropolitics often arises in regions where countries or regions share transboundary rivers, lakes, or aquifers, leading to potential conflicts over water rights, access, and management.

Impact of Water Scarcity on Food Prices: Water scarcity can lead to a reduction in agricultural productivity, especially in water-intensive crops, leading to a decrease in food supply. This in turn drives up food prices. In many regions, water scarcity is a key driver of food insecurity, affecting both local food systems and global markets.

Industrial Water Use: Refers to the consumption of water by industries for various purposes such as cooling, processing, manufacturing, and cleaning. Industrial water use is a significant part of global water demand, and the efficient management of this water is essential to avoid unnecessary consumption and reduce the environmental impact.

Integrated Water Resources Management (IWRM): A process that promotes the coordinated development and management of water, land, and related resources to maximize economic and social welfare in an equitable and sustainable manner. IWRM focuses on balancing different water needs, promoting stakeholder participation, and ensuring that water systems are resilient to changing environmental conditions.

Irrigation Efficiency: The effectiveness with which irrigation systems deliver water to crops, reducing water wastage and ensuring optimal crop growth. Improving irrigation efficiency can help conserve water, increase agricultural productivity, and mitigate the impacts of water scarcity, especially in regions heavily dependent on irrigation for food production.

Jordan River Tensions: Refers to the political and diplomatic conflicts between Israel, Jordan, and Palestine over access to and control of the Jordan River, a crucial water source in the Middle East. Disputes over water allocation have been a significant issue in the region, affecting not only local populations but also broader geopolitical relations.

Low Rainfall: A climatic condition where an area receives insufficient precipitation, leading to drought and water scarcity. Low rainfall is a major challenge in arid and semi-arid regions and exacerbates water stress, impacting agriculture, drinking water supply, and overall economic activities.

Marine Ecosystems: Ecosystems in oceans, seas, and coastal areas, which include marine life forms such as fish, corals, and plankton, and play a vital role in regulating the Earth's climate, carbon cycle, and water systems. Healthy marine ecosystems are essential for food security, biodiversity, and the livelihoods of millions of people, particularly in coastal regions.

MENA Region Water Issues: Refers to the water scarcity challenges faced by countries in the Middle East and North Africa (MENA) region, which is one of the driest regions in the world. These issues include over-extraction of water resources, poor water management, political conflicts over shared water resources, and the impact of climate change on water availability.

Middle East Water Crisis: A regional crisis driven by water scarcity, political tensions, and conflict over shared water resources, such as the Nile, Jordan, and Tigris-Euphrates Rivers. The Middle East's dependence on limited freshwater sources and the growing population make water a highly contested resource, exacerbated by environmental stressors like climate change.

Mining and Water Pollution: The process of mining can lead to the contamination of nearby water bodies through the release of toxic substances, such as heavy metals, chemicals, and sediments. This pollution can degrade water quality, harm aquatic life, and affect local communities that depend on the water for drinking, agriculture, and sanitation.

Modernization Needs: Refers to the necessity of upgrading water infrastructure, technologies, and management practices to meet contemporary challenges. This includes implementing efficient irrigation systems, improving water treatment and desalination technologies, and modernizing water governance to ensure equitable and sustainable water use.

Monsoon Variability: Refers to changes in the timing, intensity, and duration of the monsoon season, which can significantly affect water availability in regions dependent on seasonal rains, particularly in South Asia. Variability can lead to either floods or droughts, both of which have substantial impacts on agriculture, water supply, and livelihoods.

Niger River Basin Disputes: Conflicts and tensions between countries in West Africa (such as Niger, Mali, and Nigeria) over the allocation and use of water from the Niger River. These disputes are influenced by factors like population growth, agricultural needs, and hydroelectric development, and they threaten regional cooperation and sustainable water management.

Nile River Disputes: Refers to ongoing conflicts between Egypt, Sudan, Ethiopia, and other countries over the allocation of water from the Nile River. The construction of large dams, such as the Grand Ethiopian Renaissance Dam (GERD), has intensified tensions, with downstream countries (Egypt and Sudan) fearing reduced water flow for irrigation and drinking.

Regional Water Cooperation: Involves the collaborative management of shared water resources between neighboring regions, countries, or states. This cooperation can be formalized through treaties, agreements, or joint initiatives, aiming to ensure equitable water allocation, sustainable resource use, and the prevention of conflicts over water. Regional cooperation is crucial in areas with transboundary rivers or aquifers, where water use in one area can impact others.

Renewable Freshwater: Refers to freshwater resources that are replenished through natural processes, such as rainfall, snowmelt, or groundwater recharge, on a regular basis. This term distinguishes between renewable freshwater and non-renewable resources like fossil water (aquifers that do not naturally recharge). Sustainable management of renewable freshwater is essential to maintaining water availability in the long term.

Resilient Infrastructure: Infrastructure that is designed to withstand, adapt to, and recover from various environmental stresses and climate change impacts. In the context of water, resilient infrastructure includes systems for water supply, wastewater treatment, and flood management that are robust, flexible, and can continue to function effectively during extreme events such as droughts, floods, and storms.

River Basin Management: The integrated management of the entire river system, including the river, its tributaries, and the surrounding watershed. Effective river basin management considers water quality, ecosystem health, flood control, and equitable distribution of water resources among users. It involves coordination across administrative and geographical boundaries, incorporating scientific research, policy-making, and stakeholder engagement.

Rural Livelihoods and Water: The relationship between water availability and the livelihoods of rural communities, particularly those who depend on agriculture, livestock, and small-scale industries for their income. Access to reliable and clean water is essential for crop irrigation, livestock watering, and daily needs, with water scarcity posing significant threats to food security, health, and income generation in rural areas.

Saudi Arabia Water Issues: Saudi Arabia faces significant water challenges due to its arid climate, lack of natural freshwater resources, and reliance on desalination for its water supply. The country has implemented a range of strategies, including desalination, wastewater recycling, and groundwater extraction, to meet its water demand, but these solutions come with environmental, economic, and sustainability concerns, particularly regarding the depletion of groundwater and the energy-intensive nature of desalination.

Seasonal Water Scarcity: Occurs when water resources are not evenly available throughout the year, typically due to seasonal variations in rainfall, temperature, or snowmelt. Regions experiencing seasonal water scarcity often face water shortages during dry seasons or periods of drought, affecting agriculture, drinking water supplies, and ecosystem health. Managing seasonal variability requires adaptive water management strategies.

Shared Water Resources: Water resources (such as rivers, lakes, and aquifers) that are shared by two or more countries, regions, or communities. These shared resources require cooperative management to ensure equitable access, prevent conflicts, and promote sustainable use. Disputes often arise over the allocation of shared water resources, especially in regions where water is scarce or politically sensitive.

Social Impacts of Water Shortages: Water shortages can have profound social consequences, including increased poverty, migration, gender inequality, health risks, and social unrest. In many communities, women and children bear the brunt of water scarcity, as they are typically responsible for water collection, often traveling long distances for water. Access to clean water is also critical for maintaining public health and social stability.

Socioeconomic Water Policies: Policies that address the relationship between water management and socioeconomic factors such as poverty, inequality, and access to basic services. These policies aim to balance water demand across different sectors (agriculture, industry, domestic use) and ensure that water resources are distributed equitably, considering the needs of marginalized groups and communities. Effective socioeconomic policies promote both economic growth and social well-being while maintaining sustainability.

South Africa Desalination: South Africa has turned to desalination as part of its strategy to address water scarcity, particularly in coastal areas like Cape Town. Desalination involves the process of removing salt and other impurities from seawater to produce freshwater. While it can provide a reliable source of water in drought-prone regions, desalination is energy-intensive and expensive, making it less sustainable in the long term without a renewable energy source.

Southeast Asia Water Security: Refers to the region's ability to ensure reliable access to sufficient and safe water for its population. Southeast Asia faces challenges such as water scarcity, pollution, and the impacts of climate change, including flooding and droughts. Water security in the region requires coordinated management of shared water resources, improved infrastructure, and sustainable practices in agriculture and industry.

Sub-Saharan Africa Water Crisis: Sub-Saharan Africa faces critical water challenges due to factors such as climate variability, limited infrastructure, population growth, and inadequate water management. The crisis includes both physical water scarcity (in arid and semi-arid regions) and economic water scarcity (where water is available but not accessible to many people). Addressing the water crisis requires investments in infrastructure, policies to improve water access, and measures to protect water quality.

Supply-Side Water Management: Refers to strategies aimed at increasing the supply of water to meet demand, typically through measures such as the construction of dams, reservoirs, desalination plants, or improving the efficiency of water systems. Supply-side management also involves water storage, inter-basin transfers, and infrastructure development to ensure reliable water delivery, especially in areas facing increasing demand or climate-induced variability.

Sustainable Water Management: The practice of managing water resources in a way that meets current and future needs without compromising the ability of future generations to meet their own water needs. Sustainable water management involves a balance of supply and demand, conservation, pollution control, and ecosystem protection. It also includes integrating water management into broader environmental, social, and economic policies to ensure long-term water security.

Technological Innovations: Refers to the development and application of new technologies to improve water management, increase efficiency, and address water scarcity challenges. Innovations include smart water meters, desalination technologies, wastewater treatment improvements, leak detection systems, and advanced irrigation systems. These technologies can help optimize water use, reduce waste, and provide new sources of clean water.

Transboundary Water Conflicts: Conflicts that arise between countries or regions sharing a common water resource, such as a river or aquifer. These conflicts often stem from disputes over water allocation, usage rights, and environmental protection. Managing transboundary waters requires diplomatic negotiations, cooperation, and international agreements to ensure equitable and sustainable use of shared resources.

Transboundary Water Governance: The process of managing shared water resources across political boundaries, involving multiple stakeholders from different countries or regions. Effective governance is critical to resolving conflicts, ensuring equitable distribution, protecting ecosystems, and fostering cooperation. Transboundary water governance typically involves treaties, institutions, and collaborative frameworks for managing joint water resources.

Turkey's Dams Impact: Turkey has constructed numerous dams, particularly on the Tigris and Euphrates rivers, which are shared with neighboring countries like Syria and Iraq. These dams provide hydroelectric power, irrigation, and flood control but have significant geopolitical and environmental impacts, including reduced water flow downstream, ecological degradation, and tensions with neighboring nations over water rights and access.

UAE Water Management: Water management in the United Arab Emirates (UAE) focuses on addressing the country's water scarcity issues due to its arid climate and limited natural freshwater resources. The UAE relies heavily on desalination (producing freshwater from seawater) and groundwater extraction. Water management strategies in the UAE include technological innovations, water conservation policies, wastewater recycling, and the development of infrastructure like water storage systems and distribution networks to meet the demands of both urban and agricultural sectors.

Uneven Water Distribution: Refers to the disparity in water availability across different regions or communities. Water resources are often unevenly distributed due to geographical factors (such as proximity to rivers, rainfall patterns, and terrain), leading to significant differences in water access. Some regions may have abundant freshwater supplies, while others face severe scarcity. Uneven water distribution can lead to social, economic, and political conflicts, particularly in areas where water is a critical resource.

Urban Water Management: The process of managing water resources within urban areas to ensure a reliable supply for domestic, industrial, and public uses. Urban water management includes the planning and development of infrastructure like water treatment plants, pipelines, sewage systems, and stormwater management. It also focuses on the conservation of water, the reduction of waste, and the promotion of sustainable water practices to support growing urban populations and prevent water-related issues such as shortages or flooding.

Urbanization Strain: Refers to the pressure placed on water resources and infrastructure due to rapid urbanization. As cities grow, the demand for water increases, often outpacing the capacity of existing water supply systems. Urbanization strain can lead to water shortages, poor water quality, and inefficient water distribution, necessitating the development of robust water management policies and investment in water infrastructure to cope with increasing demands.

Virtual Water Trade: The concept of "virtual water" refers to the hidden water used to produce goods and services, especially in agricultural production. Virtual water trade involves the exchange of goods that require significant amounts of water to produce, such as crops or meat. Countries or regions with limited freshwater resources may import goods with high virtual water content to reduce pressure on their own water supplies. Conversely, water-abundant areas may export these water-intensive goods.

Water Abundance Challenges: Refers to the difficulties faced by regions with an abundance of water in terms of managing and distributing it efficiently. While water-rich areas may not face immediate shortages, they may struggle with issues such as over-extraction, pollution, and the need to manage water quality and ecosystem health. Additionally, water abundance in some regions does not always equate to equitable access for all populations, leading to potential challenges in distribution and equity.

Water Allocation: The process of distributing water among various sectors (such as domestic, agricultural, industrial) and across different regions or countries. Water allocation is often influenced by factors such as population size, water availability, and economic needs. Efficient water allocation aims to meet the demands of all stakeholders while maintaining the sustainability of water resources and ensuring that vulnerable populations have access to clean water.

Water Cooperation: Refers to the collaborative efforts between countries, regions, or stakeholders to manage shared water resources, especially in transboundary river basins or aquifers. Water cooperation can involve agreements or treaties on water allocation, joint infrastructure projects, and environmental protection measures. The goal is to ensure equitable and sustainable water use, reduce conflicts, and foster regional stability.

Water Crisis Mitigation: Strategies and actions designed to address and reduce the impacts of water crises, which may involve water scarcity, pollution, or poor distribution. Mitigation efforts include improving water efficiency, investing in water infrastructure, promoting water conservation practices, developing alternative water sources like desalination or wastewater recycling, and implementing policies that ensure equitable access to clean water.

Water Demand: Refers to the total amount of water required for various human, agricultural, and industrial activities. Water demand can vary significantly depending on factors such as population size, economic activities, climate conditions, and technological advancements. Managing water demand effectively is crucial to prevent over-extraction and ensure sustainable use of available resources.

Water Distribution Challenges: The difficulties in ensuring equitable and efficient distribution of water to all users, especially in areas with limited water resources. Challenges may include aging infrastructure, unequal access to water in rural versus urban areas, inefficient water systems, and political or social barriers that prevent fair distribution. Addressing these challenges requires investment in infrastructure, improved management practices, and policy reforms.

Water Ecosystem Resilience: The ability of aquatic ecosystems (such as rivers, lakes, and wetlands) to recover from disturbances, adapt to changing conditions, and continue providing essential services such as water purification, flood regulation, and habitat for biodiversity. Resilient water ecosystems are essential for maintaining water quality and quantity, as well as supporting the livelihoods of communities that depend on these resources.

Water Ethics: A branch of ethics concerned with the moral principles surrounding the use and management of water resources. Water ethics addresses issues such as equitable access to water, the right to clean water, the moral responsibility to conserve water, and the environmental impact of water extraction and pollution. It also involves considering the needs of future generations and balancing human, ecological, and economic interests.

Water Footprint: Refers to the total volume of water used to produce goods and services, including both direct consumption (e.g., drinking water) and indirect use (e.g., water used in the production of food, clothing, and energy). The water footprint is used to assess the environmental impact of water consumption and to promote sustainable water practices by understanding the hidden water costs of products and activities.

Water Governance: The processes, institutions, and policies that determine how water resources are managed, allocated, and protected. Effective water governance ensures that water is used equitably, efficiently, and sustainably, and it involves decision-making at local, national, and international levels. Water governance includes legal frameworks, stakeholder engagement, and monitoring and enforcement mechanisms.

Water Harvesting: The practice of collecting and storing rainwater or runoff for later use. Water harvesting techniques can include the construction of reservoirs, cisterns, or rainwater collection systems that capture water from rooftops or other surfaces. This method is particularly useful in areas with limited access to freshwater and can help improve water availability for drinking, irrigation, and sanitation.

Water Inequality: Refers to the unequal access to water resources, with disparities based on geographic location, socioeconomic status, and political power. Water inequality can manifest in urban-rural divides, where rural areas often have limited access to clean water compared to urban centers, or in regions where water resources are unevenly distributed or polluted, preventing certain communities from obtaining safe water.

Water Infrastructure: The physical systems and structures needed to supply, store, and manage water resources. This includes pipes, treatment plants, dams, reservoirs, and distribution networks that transport water to users. Robust water infrastructure is essential for ensuring a reliable water supply, improving water quality, and managing water use efficiently.

Water Management Innovation: Refers to new ideas, technologies, and approaches to improve the efficiency, sustainability, and equity of water management. Innovations in water management include smart water meters, advanced irrigation systems, water recycling technologies, and digital tools for monitoring and managing water usage. These innovations aim to address challenges such as water scarcity, pollution, and infrastructure deficiencies.

Water Over-extraction: The practice of extracting water from natural sources at a rate faster than it can be replenished. Over-extraction depletes aquifers, reduces river flows, and can lead to long-term environmental damage, including the collapse of ecosystems, land subsidence, and saltwater intrusion in coastal areas. Sustainable water management aims to prevent over-extraction by balancing supply and demand.

Water Pollution: The contamination of water resources (such as rivers, lakes, and groundwater) with harmful substances like chemicals, pathogens, or waste. Water pollution can result from industrial discharge, agricultural runoff, untreated sewage, and urban waste. It poses significant risks to human health, aquatic ecosystems, and the environment, necessitating pollution control measures and water treatment solutions.

Water Quality Degradation: The deterioration of water quality due to pollutants, such as chemicals, nutrients, heavy metals, or pathogens, which can make water unsafe for consumption and disrupt aquatic ecosystems. Water quality degradation is a major concern for public health and biodiversity, requiring regulation, monitoring, and effective treatment technologies to maintain clean water resources.

Water Recycling: The process of reusing treated wastewater for various purposes, such as irrigation, industrial processes, and even potable water supply. Water recycling helps reduce demand for fresh water and is a key strategy for addressing water scarcity in regions where freshwater resources are limited or overused.

Water Scarcity: The condition where water demand exceeds available supply, either due to physical scarcity (where water resources are insufficient to meet the needs of a population) or economic scarcity (where access to water is hindered by poverty or inadequate infrastructure). Water scarcity can lead to food insecurity, health problems, and social conflicts, and addressing it requires sustainable water management practices.

Water Security Strategies: Approaches to ensure the availability, access, and quality of water for all people, sectors, and ecosystems, both now and in the future. Water security strategies involve improving water governance, investing in infrastructure, promoting water conservation, protecting water ecosystems, and adapting to climate change impacts. These strategies aim to prevent water shortages and ensure sustainable access to safe water.

Water Stress Indicators: Measures used to assess the level of stress on water resources within a given region. Indicators of water stress include the ratio of water withdrawals to available freshwater resources, water scarcity, and the frequency of droughts

Major topics discussed in the book.

Water Availability and Distribution: Examines the global patterns of freshwater distribution, highlighting disparities between regions with extreme scarcity and those with relative abundance.

Climate Change and Water Resources: Analyzes the impact of climate change on water availability, including shifts in precipitation patterns, increased drought frequency, and melting glaciers.

Water Stress and Scarcity Indicators: Discusses various metrics and indicators used to measure water stress, scarcity, and security, such as the Water Stress Index and the Falkenmark Indicator.

Groundwater Depletion: Explores the over-extraction of groundwater resources, focusing on regions experiencing significant aquifer depletion and the consequences for agriculture and communities.

Water Pollution and Contamination: Covers the sources of water pollution, including industrial waste, agricultural runoff, and untreated sewage, and their effects on freshwater ecosystems and human health.

Integrated Water Resources Management (IWRM): Evaluates strategies for sustainable water management, emphasizing the principles of IWRM to balance social, economic, and environmental needs.

Desalination and Water Reuse Technologies: Explores innovative technologies for addressing water scarcity, such as desalination plants and wastewater recycling, and their potential to enhance water security.

Water-Energy Nexus: Discusses the interdependence between water and energy systems, highlighting the challenges of water use in energy production and energy use in water treatment.

Transboundary Water Conflicts and Cooperation: Analyzes case studies of shared water resources, focusing on international water treaties, disputes, and collaborative efforts to manage transboundary waters.

Policies and Governance for Water Security: Reviews global and regional policies aimed at ensuring water security, including regulatory frameworks, water pricing, and **community**-based management approaches.

About the Author

Dr. Medani P. Bhandari is a widely recognized humanitarian, professor, author, and editor who has significantly contributed to social and environmental sciences. With expertise in social equity, inequality, inclusion, innovation, feminism, sustainability, climate change, and social and environmental policies and management, Dr. Bhandari has dedicated his career to these important areas. In addition to his extensive scholarly work, Dr. Bhandari is a poet, essayist, and advocate for the environment and social causes. His expertise expands to global and international environmental politics and justice, climate change, sustainable development, public/social policy, the non-profit sector, renewable energy, nature, culture, and power. He has gained field experience across diverse regions, including Asia, Africa, North America, Western Europe, Australia, Japan, and the Middle East. Dr. Bhandari has authored and co-edited several books, including Green Web-II: Standards and Perspectives from the IUCN, 2nd Edition, Getting the Climate Science Facts Right: The Role of the IPCC, Reducing Inequalities Towards Sustainable Development Goals: Multilevel Approach, Educational Transformation, Economic Inequality – Trends, Traps, and Trade-offs, Social Inequality as a Global Challenge, The Unbeatable Challenges of Inequality, Perspectives on Sociological Theories, Methodological Debates, and Organizational Sociology, The Past, Present, and Future Directions of Migration/Immigration, Theories, and Practices of Global Journeys, Transforming Futures – Navigating Society, Environment, and Governance, Live and Let Others Live – In Reference to Sustainability and Environment Conservation, Navigating Sustainability: A Comparative Analysis of Micro vs. Macro Strategies in the least Developed Countries, The Nexus of Climate Change and Land-use – Global Scenario with Reference to Nepal, Women and Society etc.. Social Inequality- Past, Present, and Future (in which 15 African Countries' case studies are included), etc. Furthermore, Dr. Bhandari's creative writing encompasses a vast collection of poems and essays. He has published three volumes of poetry in collaboration with

Prajita Bhandari. Dr. Bhandari\'s multifaceted contributions to academia, humanitarianism, and creative writing have solidified his reputation as an esteemed figure in social and environmental sciences, profoundly impacting social equity, sustainability, climate change, and beyond. Professor Bhandari holds five master's degrees in social and environmental sciences and a PhD in sociology. For 45 years, he has been actively involved in environmental conservation and humanitarian efforts through nonprofit and academic platforms. Guided by the philosophy that all living and non-living beings are interconnected ("Vasudhaiva Kutumbakam") and the principle of "Live and Let Others Live," his work reflects a deep commitment to the well-being of our planet and its inhabitants.

He serves as an International Program Director at the New York-based NGO Atlantic State Legal Foundation in Syracuse, NY; a professor at Akamai University, USA, and Sumy State University, Ukraine; and an advisor to Gandaki University, Nepal. He is also the UNEP Civil Society Regional Facilitator for North America and the Lead Author of UNEP GEO 7. He is also the editor-in-chief of Strategic Planning for Energy and the Environment, the books series editor at the River Publisher, and the books series editor at Nova Publishers.

For Product Safety Concerns and Information please contact our
EU representative GPSR@taylorandfrancis.com Taylor & Francis
Verlag GmbH, Kaufingerstraße 24, 80331 München, Germany